Ben Stacy Jerrik (Ed.)

Bullet time

Ben Stacy Jerrik (Ed.)

Bullet time

Simulation, Photography, Visual effects

Part Press

Imprint

Publisher:
Part Press is a trademark of
International Book Market Service Ltd., 17 Rue Meldrum, Beau Bassin, 1713-01 Mauritius
Email: info@bookmarketservice.com
Website: www.bookmarketservice.com

Published in 2012

Printed in: U.S.A., U.K., Germany. This book was not produced in Mauritius.

ISBN: 978-613-5-61835-8

Contents

Articles

References

Bullet_time

Bullet time (also known as **frozen time, the big freeze, dead time, flow motion**, or **time slice**)[1] is a special and visual effect that refers to a digitally enhanced simulation of variable-speed (i.e. slow motion, time-lapse, etc.) photography used in films, broadcast advertisements, and video games. It is characterized both by its extreme transformation of time (slow enough to show normally imperceptible and unfilmable events, such as flying bullets) and space (by way of the ability of the camera angle—the audience's point-of-view—to move around the scene at a normal speed while events are slowed). This is almost impossible with conventional slow-motion, as the physical camera would have to move impossibly fast; the concept implies that only a "virtual camera", often illustrated within the confines of a computer-generated environment such as a virtual world or virtual reality, would be capable of "filming" bullet-time types of moments. Technical and historical variations of this effect have been referred to as time slicing, view morphing, slow-mo, *temps mort*, and virtual cinematography.

Technology

The bullet time effect was originally achieved photographically by a set of still cameras surrounding the subject. The cameras are fired sequentially, or all at the same time, depending on the desired effect. Single frames from each camera are then arranged and displayed consecutively to produce an orbiting viewpoint of an action frozen in time or as hyper-slow-motion. This technique suggests the limitless perspectives and variable frame rates possible with a virtual camera. However, if the still array process is done with real cameras, it is often limited to assigned paths.

For many years, it has been possible to use computer vision techniques to capture scenes and render images of novel viewpoints sufficient for bullet time type effects. More recently, these have been formalized into what is becoming known as free viewpoint television (FTV). At the time of *The Matrix*, FTV was not a fully mature technology. FTV is effectively the live action version of bullet time, without the slow motion.

In *The Matrix*, the camera path was pre-designed using computer-generated visualizations as a guide. Cameras were arranged, behind a green or blue screen, on a track and aligned through a laser targeting system, forming a complex curve through space. The cameras were then triggered at extremely close intervals, so the action continued to unfold, in extreme slow-motion, while the viewpoint moved. Additionally, the individual frames were scanned for computer processing. Using sophisticated interpolation software, extra frames could be inserted to slow down the action further and improve the fluidity of the movement (especially the frame rate of the images); frames could also be dropped to speed up the action. This approach provides greater flexibility than a purely photographic one. The same effect can also be produced using pure CGI, motion capture and universal capture. It is thought that the opening sequence from the late 1960s *Speed Racer* cartoons partially inspired The Wachowskis to incorporate the bullet time effect into *The Matrix*.

History

The technique of using a group of still cameras to freeze motion occurred before the invention of cinema itself. It dates back to the 19th-century experiments by Eadweard Muybridge, who analyzed the motion of a galloping horse by using a line of cameras to photograph the animal as it ran past.[1] Eadweard Muybridge used still cameras placed along a racetrack, and each camera was actuated by a taut string stretched across the track; as the horse galloped past, the camera shutters snapped, taking one frame at a time. (The original intent was to settle a debate the governor of California had engaged in, as to whether all four of the animal's legs would leave the ground when galloping.) Muybridge later assembled the pictures into a rudimentary animation, by placing them on a glass disk which he spun in front of a light source. His zoopraxiscope may have been an inspiration for Thomas Edison to explore the idea of motion pictures.[2]

Muybridge horse photos

Muybridge also took photos of actions from many angles at the same instant in time, to study how the human body went up stairs, for example. This is the effect used in *The Matrix*, and other movies, and is achieved roughly the same way as Muybridge set up his shots. In effect, however, Muybridge had achieved the aesthetic opposite to *The Matrix*'s bullet-time sequences, since his studies lacked the dimensionality of the later developments. A debt may also be owed to MIT professor Doc Edgerton, who, in the 1940s, captured now-iconic photos of bullets using xenon strobe lights to "freeze" motion.[3]

Long before the emergence of a technology permitting a live-action application, bullet-time as a concept was frequently developed in cel animation. One of the earliest examples is the shot at the end of the title sequence for the 1966 Japanese anime series *Speed Racer*: as Speed leaps from the Mach Five, he freezes in mid-jump, and then the camera does an arc shot from front to sideways.

In 1980, Tim Macmillan started producing pioneering video work[4] in this field while studying for a BA at the (then named) Bath Academy of Art using 16mm film arranged in a progressing sequence of pinhole cameras.

The first music video to use bullet-time was "Midnight Mover", a 1985 Accept video.[5] A variation on bullet-time came into use in the mid-1990s, first in a series of commercials to sell products for Gap and Lexus, then in feature films such as *Lost in Space*, and later in *The Matrix*.[1] Bullet time became popularized when John Gaeta and team expanded it temporally and into the digital arena through the incorporation of frame interpolation and image-based CGI within the film *The Matrix* (1999) and through view-morphing techniques pioneered by BUF Compagnie in music videos by Michel Gondry and commercials for, among others, The Gap. Gaeta said of *The Matrix*'s use of the effect:

> For artistic inspiration for bullet time, I would credit Otomo Katsuhiro, who co-wrote and directed *Akira*, which definitely blew me away, along with director Michel Gondry. His music videos experimented with a different type of technique called view-morphing and it was just part of the beginning of uncovering the creative approaches toward using still cameras for special effects. Our technique was significantly different because we built it to move around objects that were themselves in motion, and we were also able to create slow-motion events that 'virtual cameras' could move around – rather than the static action in Gondry's music videos with limited camera moves.[6]

In popular film culture, *Blade* (1998) was the first film to illustrate super-human reflexes dodging bullets in a scene in a park when the character of Blade fires at Stephen Dorff's character, Deacon Frost. Computer generated bullets are also used to illustrate the concept and slow motion camera work.

John Woo is also famous for incorporating slow motion shoot outs in his movies, such as *A Better Tomorrow* and *Hard Boiled*. These films would later influence media such as *Max Payne* and *The Matrix* series.

In 2003, bullet time evolved further through *The Matrix Reloaded* and *The Matrix Revolutions* with the introduction of high-definition computer-generated approaches like virtual cinematography and universal capture. Virtual elements within the Matrix Trilogy utilized state-of-the-art image-based computer rendering techniques pioneered in Paul Debevec's 1997 film [7] and custom evolved for the Matrix by George Borshukov, an early collaborator of Debevec.

Following *The Matrix*, bullet time has featured as a key gameplay mechanic in various video games, which typically involve players having the ability to slow down time during certain actions, allowing them to easily aim weapons. The first video game to use this style of gameplay was Remedy Entertainment's 2001 video game, *Max Payne*. Bullet time was used for the first time in a live music environment in October 2009 for Creed's live DVD *Creed Live*.[8]

References

[1] Argy, Stephanie (21 January 2001). "Frozen f/x still in action: There's less love for morph" (http://www.variety.com/article/VR1117792382?refCatId=2400). Variety.com. . Retrieved 3 April 2012.

[2] Hendricks, Gordon (1961), *The Edison Motion Picture Myth*, Berkeley, California: University of California Press

[3] "High Speed Camera « Harold "Doc" Edgerton" (http://edgerton-digital-collections.org/techniques/high-speed-photography). 2009-11-28. . Retrieved 2009-11-28.

[4] Video of Tim Macmillan Early Work 1980 - 1994 (http://vimeo.com/6165108) on Vimeo

[5] ACCEPT Remembered - Discography - Metal Heart (http://web.archive.org/web/20020215044049/wolfhoffmann.com/html/metalheart.html)

[6] "200 Things That Rocked Our World: Bullet Time". *Empire* (EMAP) (200): 136. February 2006.

[7] "The Campanile Movie" (http://ict.debevec.org/~debevec/Campanile/). Paul Debevec. Updated November 2007. . Retrieved September 2009.

[8] "Creed Announce First Live DVD" (http://www.guitarworld.com/article/creed_announce_first_live_dvd). Guitar World. November 24, 2009. .

External links

- Bullet Time (http://tvtropes.org/pmwiki/pmwiki.php/Main/BulletTime) at TV Tropes
- Frozen moment might become the morph of the late '90s (http://www.variety.com/article/VR1117792382?refCatId=2400)

Simulation

Simulation is the imitation of the operation of a real-world process or system over time.[1] The act of simulating something first requires that a model be developed; this model represents the key characteristics or behaviors of the selected physical or abstract system or process. The model represents the system itself, whereas the simulation represents the operation of the system over time.

Wooden mechanical horse simulator during WWI.

Simulation is used in many contexts, such as simulation of technology for performance optimization, safety engineering, testing, training, education, and video games. Training simulators include flight simulators for training aircraft pilots to provide them with a lifelike experience. Simulation is also used with scientific modelling of natural systems or human systems to gain insight into their functioning.[2] Simulation can be used to show the eventual real effects of alternative conditions and courses of action. Simulation is also used when the real system cannot be engaged, because it may not be accessible, or it may be dangerous or unacceptable to engage, or it is being designed but not yet built, or it may simply not exist.[3]

Key issues in simulation include acquisition of valid source information about the relevant selection of key characteristics and behaviours, the use of simplifying approximations and assumptions within the simulation, and fidelity and validity of the simulation outcomes.

Classification and terminology

Historically, simulations used in different fields developed largely independently, but 20th century studies of Systems theory and Cybernetics combined with spreading use of computers across all those fields have led to some unification and a more systematic view of the concept.

Human-in-the-loop simulation of outer space.

Physical simulation refers to simulation in which physical objects are substituted for the real thing (some circles[4] use the term for computer simulations modelling selected laws of physics, but this article doesn't). These physical objects are often chosen because they are smaller or cheaper than the actual object or system.

Interactive simulation is a special kind of physical simulation, often referred to as a *human in the loop* simulation, in which physical simulations include human operators, such as in a flight simulator or a driving simulator.

Human in the loop simulations can include a computer simulation as a so-called *synthetic environment*.[5]

Computer simulation

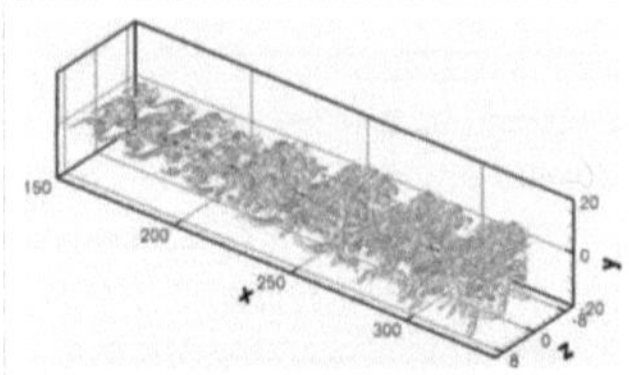
Visualization of a direct numerical simulation model.

A computer simulation (or "sim") is an attempt to model a real-life or hypothetical situation on a computer so that it can be studied to see how the system works. By changing variables in the simulation, predictions may be made about the behaviour of the system. It is a tool to virtually investigate the behaviour of the system under study.[1]

Computer simulation has become a useful part of modeling many natural systems in physics, chemistry and biology,[6] and human systems in economics and social science (the computational sociology) as well as in engineering to gain insight into the operation of those systems. A good example of the usefulness of using computers to simulate can be found in the field of network traffic simulation. In such simulations, the model behaviour will change each simulation according to the set of initial parameters assumed for the environment.

Traditionally, the formal modeling of systems has been via a mathematical model, which attempts to find analytical solutions enabling the prediction of the behaviour of the system from a set of parameters and initial conditions. Computer simulation is often used as an adjunct to, or substitution for, modeling systems for which simple closed form analytic solutions are not possible. There are many different types of computer simulation, the common feature they all share is the attempt to generate a sample of representative scenarios for a model in which a complete enumeration of all possible states would be prohibitive or impossible.

Several software packages exist for running computer-based simulation modeling (e.g. Monte Carlo simulation, stochastic modeling, multimethod modeling) that makes all the modeling almost effortless.

Modern usage of the term "computer simulation" may encompass virtually any computer-based representation.

Computer science

In computer science, simulation has some specialized meanings: Alan Turing used the term "simulation" to refer to what happens when a universal machine executes a state transition table (in modern terminology, a computer runs a program) that describes the state transitions, inputs and outputs of a subject discrete-state machine. The computer simulates the subject machine. Accordingly, in theoretical computer science the term *simulation* is a relation between state transition systems, useful in the study of operational semantics.

Less theoretically, an interesting application of computer simulation is to simulate computers using computers. In computer architecture, a type of simulator, typically called an *emulator*, is often used to execute a program that has to run on some inconvenient type of computer (for example, a newly designed computer that has not yet been built or an obsolete computer that is no longer available), or in a tightly controlled testing environment (see *Computer architecture simulator* and *Platform virtualization*). For example, simulators have been used to debug a microprogram or sometimes commercial application programs, before the program is downloaded to the target machine. Since the operation of the computer is simulated, all of the information about the computer's operation is directly available to the programmer, and the speed and execution of the simulation can be varied at will.

Simulators may also be used to interpret fault trees, or test VLSI logic designs before they are constructed. Symbolic simulation uses variables to stand for unknown values.

In the field of optimization, simulations of physical processes are often used in conjunction with evolutionary computation to optimize control strategies...

Simulation in education and training

Simulation is extensively used for educational purposes. It is frequently used by way of adaptive hypermedia.

Simulation is often used in the training of civilian and military personnel.[7] This usually occurs when it is prohibitively expensive or simply too dangerous to allow trainees to use the real equipment in the real world. In such situations they will spend time learning valuable lessons in a "safe" virtual environment yet living a lifelike experience (or at least it is the goal). Often the convenience is to permit mistakes during training for a safety-critical system. For example, in simSchool [8] teachers practice classroom management and teaching techniques on simulated students, which avoids "learning on the job" that can damage real students. There is a distinction, though, between simulations used for training and Instructional simulation.

Training simulations typically come in one of three categories:[9]

- "live" simulation (where actual players use genuine systems in a real environment);
- "virtual" simulation (where actual players use simulated systems in a synthetic environment [5]), or
- "constructive" simulation (where simulated players use simulated systems in a synthetic environment). Constructive simulation is often referred to as "wargaming" since it bears some resemblance to table-top war games in which players command armies of soldiers and equipment that move around a board.

In standardized tests, "live" simulations are sometimes called "high-fidelity", producing "samples of likely performance", as opposed to "low-fidelity", "pencil-and-paper" simulations producing only "signs of possible performance",[10] but the distinction between high, moderate and low fidelity remains relative, depending on the context of a particular comparison.

Simulations in education are somewhat like training simulations. They focus on specific tasks. The term 'microworld' is used to refer to educational simulations which model some abstract concept rather than simulating a realistic object or environment, or in some cases model a real world environment in a simplistic way so as to help a learner develop an understanding of the key concepts. Normally, a user can create some sort of construction within the microworld that will behave in a way consistent with the concepts being modeled. Seymour Papert was one of the first to advocate the value of microworlds, and the Logo programming environment developed by Papert is one of the most famous microworlds. As another example, the Global Challenge Award online STEM learning web site uses microworld simulations to teach science concepts related to global warming and the future of energy. Other projects for simulations in educations are Open Source Physics, NetSim etc.

Management games (or *business simulations*) have been finding favour in business education in recent years.[11] Business simulations that incorporate a dynamic model enable experimentation with business strategies in a risk free environment and provide a useful extension to case study discussions.

Social simulations may be used in social science classrooms to illustrate social and political processes in anthropology, economics, history, political science, or sociology courses, typically at the high school or university level. These may, for example, take the form of civics simulations, in which participants assume roles in a simulated society, or international relations simulations in which participants engage in negotiations, alliance formation, trade, diplomacy, and the use of force. Such simulations might be based on fictitious political systems, or be based on current or historical events. An example of the latter would be Barnard College's "Reacting to the Past" series of educational simulations.[12] The "Reacting to the Past" series also includes simulation games that address science education.

In recent years, there has been increasing use of social simulations for staff training in aid and development agencies. The Carana simulation, for example, was first developed by the United Nations Development Programme, and is now used in a very revised form by the World Bank for training staff to deal with fragile and conflict-affected countries.[13]

Common user interaction systems for virtual simulations

Virtual simulations represent a specific category of simulation that utilizes simulation equipment to create a simulated world for the user. Virtual simulations allow users to interact with a virtual world. Virtual worlds operate on platforms of integrated software and hardware components. In this manner, the system can accept input from the user (e.g., body tracking, voice/sound recognition, physical controllers) and produce output to the user (e.g., visual display, aural display, haptic display) .[14] Virtual Simulations use the aforementioned modes of interaction to produce a sense of immersion for the user.

Virtual simulation input hardware

There is a wide variety of input hardware available to accept user input for virtual simulations. The following list briefly describes several of them:

Motorcycle simulator of *Bienal do Automóvel* exhibition, in Belo Horizonte, Brazil.

Body tracking The motion capture method is often used to record the user's movements and translate the captured data into inputs for the virtual simulation. For example, if a user physically turns their head, the motion would be captured by the simulation hardware in some way and translated to a corresponding shift in view within the simulation.

- Capture suits and/or gloves may be used to capture movements of users body parts. The systems may have sensors incorporated inside them to sense movements of different body parts (e.g., fingers). Alternatively, these systems may have exterior tracking devices or marks that can be detected by external ultrasound, optical receivers or electromagnetic sensors. Internal inertial sensors are also available on some systems. The units may transmit data either wirelessly or through cables.
- Eye trackers can also be used to detect eye movements so that the system can determine precisely where a user is looking at any given instant.

Physical controllers Physical controllers provide input to the simulation only through direct manipulation by the user. In virtual simulations, tactile feedback from physical controllers is highly desirable in a number of simulation environments.

- Omni directional treadmills can be used to capture the users locomotion as they walk or run.
- High fidelity instrumentation such as instrument panels in virtual aircraft cockpits provides users with actual controls to raise the level of immersion. For example, pilots can use the actual global positioning system controls from the real device in a simulated cockpit to help them practice procedures with the actual device in the context of the integrated cockpit system.

Voice/sound recognition This form of interaction may be used either to interact with agents within the simulation (e.g., virtual people) or to manipulate objects in the simulation (e.g., information). Voice interaction presumably increases the level of immersion for the user.

- Users may use headsets with boom microphones, lapel microphones or the room may be equipped with strategically located microphones.

Current research into user input systems Research in future input systems hold a great deal of promise for virtual simulations. Systems such as brain-computer interfaces (BCIs)Brain-computer interface offer the ability to further increase the level of immersion for virtual simulation users. Lee, Keinrath, Scherer, Bischof, Pfurtscheller [15] proved that naïve subjects could be trained to use a BCI to navigate a virtual apartment with relative ease. Using the BCI, the authors found that subjects were able to freely navigate the virtual environment with relatively minimal effort. It is possible that these types of systems will become standard input modalities in future virtual simulation systems.simulation is a one of the part of a engineering students and also imp for main electrical students its come in form of education purpose.

Virtual simulation output hardware

There is a wide variety of output hardware available to deliver stimulus to users in virtual simulations. The following list briefly describes several of them:

Visual display Visual displays provide the visual stimulus to the user.

- Stationary displays can vary from a conventional desktop display to 360-degree wrap around screens to stereo three-dimensional screens. Conventional desktop displays can vary in size from 15 to 60+ inches. Wrap around screens are typically utilized in what is known as a Cave Automatic Virtual Environment (CAVE) Cave Automatic Virtual Environment. Stereo three-dimensional screens produce three-dimensional images either with or without special glasses—depending on the design.
- Head mounted displays (HMDs) have small displays that are mounted on headgear worn by the user. These systems are connected directly into the virtual simulation to provide the user with a more immersive experience. Weight, update rates and field of view are some of the key variables that differentiate HMDs. Naturally, heavier HMDs are undesirable as they cause fatigue over time. If the update rate is too slow, the system is unable to update the displays fast enough to correspond with a quick head turn by the user. Slower update rates tend to cause simulation sickness and disrupt the sense of immersion. Field of view or the angular extent of the world that is seen at a given moment Field of view can vary from system to system and has been found to affect the users sense of immersion.

Aural display Several different types of audio systems exist to help the user hear and localize sounds spatially. Special software can be used to produce 3D audio effects 3D audio to create the illusion that sound sources are placed within a defined three-dimensional space around the user.

- Stationary conventional speaker systems may be used provide dual or multi-channel surround sound. However, external speakers are not as effective as headphones in producing 3D audio effects.[14]
- Conventional headphones offer a portable alternative to stationary speakers. They also have the added advantages of masking real world noise and facilitate more effective 3D audio sound effects.[14]

Haptic display These displays provide sense of touch to the user Haptic technology. This type of output is sometimes referred to as force feedback.

- Tactile tile displays use different types of actuators such as inflatable bladders, vibrators, low frequency sub-woofers, pin actuators and/or thermo-actuators to produce sensations for the user.
- End effector displays can respond to users inputs with resistance and force.[14] These systems are often used in medical applications for remote surgeries that employ robotic instruments.[16]

Vestibular display These displays provide a sense of motion to the user Motion simulator. They often manifest as motion bases for virtual vehicle simulation such as driving simulators or flight simulators. Motion bases are fixed in place but use actuators to move the simulator in ways that can produce the sensations pitching, yawing or rolling. The simulators can also move in such a way as to produce a sense of acceleration on all axes (e.g., the motion base can produce the sensation of falling).

Clinical healthcare simulators

Medical simulators are increasingly being developed and deployed to teach therapeutic and diagnostic procedures as well as medical concepts and decision making to personnel in the health professions. Simulators have been developed for training procedures ranging from the basics such as blood draw, to laparoscopic surgery [17] and trauma care. They are also important to help on prototyping new devices[18] for biomedical engineering problems. Currently, simulators are applied to research and development of tools for new therapies,[19] treatments[20] and early diagnosis[21] in medicine.

Many medical simulators involve a computer connected to a plastic simulation of the relevant anatomy. Sophisticated simulators of this type employ a life size mannequin that responds to injected drugs and can be programmed to create simulations of life-threatening emergencies. In other simulations, visual components of the procedure are reproduced by computer graphics techniques, while touch-based components are reproduced by haptic feedback devices combined with physical simulation routines computed in response to the user's actions. Medical simulations of this sort will often use 3D CT or MRI scans of patient data to enhance realism. Some medical simulations are developed to be widely distributed (such as web-enabled simulations [22] that can be viewed via standard web browsers) and can be interacted with using standard computer interfaces, such as the keyboard and mouse.

Another important medical application of a simulator — although, perhaps, denoting a slightly different meaning of *simulator* — is the use of a placebo drug, a formulation that simulates the active drug in trials of drug efficacy (see Placebo (origins of technical term)).

Improving patient safety

Patient safety is a concern in the medical industry. Patients have been known to suffer injuries and even death due to management error, and lack of using best standards of care and training. According to Building a National Agenda for Simulation-Based Medical Education (Eder-Van Hook, Jackie, 2004), "A health care provider's ability to react prudently in an unexpected situation is one of the most critical factors in creating a positive outcome in medical emergency, regardless of whether it occurs on the battlefield, freeway, or hospital emergency room." simulation. Eder-Van Hook (2004) also noted that medical errors kill up to 98,000 with an estimated cost between $37 and $50 million and $17 to $29 billion for preventable adverse events dollars per year. "Deaths due to preventable adverse events exceed deaths attributable to motor vehicle accidents, breast cancer, or AIDS" Eder-Van Hook (2004). With these types of statistics it is no wonder that improving patient safety is a prevalent concern in the industry.

Innovative simulation training solutions are now being used to train medical professionals in an attempt to reduce the number of safety concerns that have adverse effects on the patients. However, according to the article Does Simulation Improve Patient Safety? Self-efficacy, Competence, Operational Performance, and Patient Safety (Nishisaki A., Keren R., and Nadkarni, V., 2007), the jury is still out. Nishisaki states that "There is good evidence that simulation training improves provider and team self-efficacy and competence on manikins. There is also good evidence that procedural simulation improves actual operational performance in clinical settings.[23] However, no evidence yet shows that crew resource management training through simulation, despite its promise, improves team operational performance at the bedside. Also, no evidence to date proves that simulation training actually improves patient outcome. Even so, confidence is growing in the validity of medical simulation as the training tool of the future." This could be because there are not enough research studies yet conducted to effectively determine the success of simulation initiatives to improve patient safety. Examples of [recently implemented] research simulations used to improve patient care [and its funding] can be found at Improving Patient Safety through Simulation Research (US Department of Human Health Services) http://www.ahrq.gov/qual/simulproj.htm.

One such attempt to improve patient safety through the use of simulations training is pediatric care to deliver just-in-time service or/and just-in-place. This training consists of 20 minutes of simulated training just before workers report to shift. It is hoped that the recentness of the training will increase the positive and reduce the negative results that have generally been associated with the procedure. The purpose of this study is to determine if just-in-time training improves patient safety and operational performance of orotracheal intubation and decrease occurrences of undesired associated events and "to test the hypothesis that high fidelity simulation may enhance the training efficacy and patient safety in simulation settings." The conclusion as reported in Abstract P38: Just-In-Time Simulation Training Improves ICU Physician Trainee Airway Resuscitation Participation without Compromising Procedural Success or Safety (Nishisaki A., 2008), were that simulation training improved resident participation in real cases; but did not sacrifice the quality of service. It could be therefore hypothesized that by increasing the number of highly trained residents through the use of simulation training, that the simulation training does in fact increase patient safety. This hypothesis would have to be researched for validation and the results may or may not generalize to other situations.

History of simulation in healthcare

The first medical simulators were simple models of human patients.[24]

Since antiquity, these representations in clay and stone were used to demonstrate clinical features of disease states and their effects on humans. Models have been found from many cultures and continents. These models have been used in some cultures (e.g., Chinese culture) as a "diagnostic" instrument, allowing women to consult male physicians while maintaining social laws of modesty. Models are used today to help students learn the anatomy of the musculoskeletal system and organ systems.[24]

Type of models

Active models

Active models that attempt to reproduce living anatomy or physiology are recent developments. The famous "Harvey" mannequin was developed at the University of Miami and is able to recreate many of the physical findings of the cardiology examination, including palpation, auscultation, and electrocardiography.[25]

Interactive models

More recently, interactive models have been developed that respond to actions taken by a student or physician.[25] Until recently, these simulations were two dimensional computer programs that acted more like a textbook than a patient. Computer simulations have the advantage of allowing a student to make judgements, and also to make errors. The process of iterative learning through assessment, evaluation, decision making, and error correction creates a much stronger learning environment than passive instruction.

Computer simulators

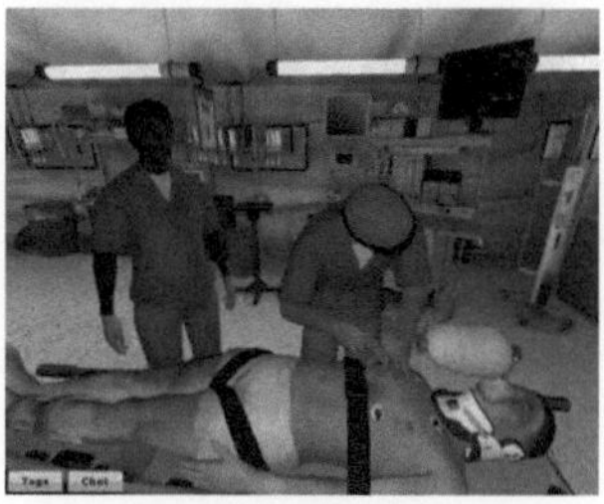
3DiTeams learner is percussing the patient's chest in virtual field hospital

Simulators have been proposed as an ideal tool for assessment of students for clinical skills.[26] For patients, "cybertherapy" can be used for sessions simulating traumatic experiecences, from fear of heights to social anxiety.[27]

Programmed patients and simulated clinical situations, including mock disaster drills, have been used extensively for education and evaluation. These "lifelike" simulations are expensive, and lack reproducibility. A fully functional "3Di" simulator would be the most specific tool available for teaching and measurement of clinical skills. Gaming platforms have been applied to create these virtual medical environments to create an interactive method for learning and application of information in a clinical context.[28] [29]

Immersive disease state simulations allow a doctor or HCP to experience what a disease actually feels like. Using sensors and transducers symptomatic effects can be delivered to a participant allowing them to experience the patients disease state.

Such a simulator meets the goals of an objective and standardized examination for clinical competence.[30] This system is superior to examinations that use "standard patients" because it permits the quantitative measurement of competence, as well as reproducing the same objective findings.[31]

Simulation in entertainment

Entertainment simulation is a term that encompasses many large and popular industries such as film, television, video games (including serious games) and rides in theme parks. Although modern simulation is thought to have its roots in training and the military, in the 20th century it also became a conduit for enterprises which were more hedonistic in nature. Advances in technology in the 1980s and 1990s caused simulation to become more widely used and it began to appear in movies such as *Jurassic Park* (1993) and in computer-based games such as Atari's *Battlezone*.

History

Early history (1940's and 50's)

The first simulation game may have been created as early as 1947 by Thomas T. Goldsmith Jr. and Estle Ray Mann. This was a straightforward game that simulated a missile being fired at a target. The curve of the missile and its speed could be adjusted using several knobs. In 1958 a computer game called "*Tennis for Two*" was created by Willy Higginbotham which simulated a tennis game between two players who could both play at the same time using hand controls and was displayed on an oscilloscope.[32] This was one of the first electronic video games to use a graphical display.

Modern simulation (1980's-present)

Advances in technology in the 1980s made the computer more affordable and more capable than they were in previous decades [33] which facilitated the rise of computer gaming. The first video game consoles released in the 1970s and early 1980s fell prey to the industry crash in 1983, but in 1985, Nintendo released the Nintendo Entertainment System (NES) which became the best selling console in video game history.[34] In the 1990s, computer games became widely popular with the release of such game as *The Sims* and *Command and Conquer* and the still increasing power of desktop computers. Today, computer simulation games such as *World of Warcraft* are played by millions of people around the world.

Computer-generated imagery was used in film to simulate objects as early as 1976, though in 1982, the film *Tron* was the first film to use computer-generated imagery for more than a couple of minutes. However, the commercial failure of the movie may have caused the industry to step away from the technology.[35] In 1993, the film *Jurassic Park* became the first popular film to use computer-generated graphics extensively, integrating the simulated dinosaurs almost seamlessly into live action scenes. This event transformed the film industry; in 1995, the film *Toy Story* was the first film to use only computer-generated images and by the new millennium computer generated graphics were the leading choice for special effects in films.[36]

Simulators have been used for entertainment since the Link Trainer in the 1930s.[37] The first modern simulator ride to open at a theme park was Disney's Star Tours in 1987 soon followed by Universal's The Funtastic World of Hanna-Barbera in 1990 which was the first ride to be done entirely with computer graphics.[38]

Examples of entertainment simulation

Computer and video games

Simulation games, as opposed to other genres of video and computer games, represent or simulate an environment accurately. Moreover, they represent the interactions between the playable characters and the environment realistically. These kinds of games are usually more complex in terms of game play.[39] Simulation games have become incredibly popular among people of all ages.[40] Popular simulation games include *SimCity*, *Tiger Woods PGA Tour* and *Virtonomics*. There are also Flight Simulation and Driving Simulation games.

Film

Computer-generated imagery is "the application of the field of 3D computer graphics to special effects". This technology is used for visual effects because they are high in quality, controllable, and can create effects that would not be feasible using any other technology either because of cost, resources or safety.[41] Computer-generated graphics can be seen in many live action movies today, especially those of the action genre. Further, computer generated imagery has almost completely supplanted hand-drawn animation in children's movies which are increasingly computer-generated only. Examples of movies that use computer-generated imagery include *Finding Nemo*, *300* and *Iron Man*.

Theme park rides

Simulator rides are the progeny of military training simulators and commercial simulators, but they are different in a fundamental way. While military training simulators react realistically to the input of the trainee in real time, ride simulators only feel like they move realistically and move according to prerecorded motion scripts.[38] One of the first simulator rides, Star Tours, which cost $32 million, used a hydraulic motion based cabin. The movement was programmed by a joystick. Today's simulator rides, such as The Amazing Adventures of Spider-Man include elements to increase the amount of immersion experienced by the riders such as: 3D imagery, physical effects (spraying water or producing scents), and movement through an environment.[42] Examples of simulation rides include Mission Space and The Simpsons Ride. There are many simulation rides at themeparks like Disney, Universal etc., Examples are Flint Stones, Earth Quake, Time Machine, King Kong.

Simulation and manufacturing

Manufacturing represents one of the most important applications of Simulation. This technique represents a valuable tool used by engineers when evaluating the effect of capital investment in equipments and physical facilities like factory plants, warehouses, and distribution centers. Simulation can be used to predict the performance of an existing or planned system and to compare alternative solutions for a particular design problem.[43]

Another important goal of manufacturing-simulations is to quantify system performance. Common measures of system performance include the following:[44]

- Throughput under average and peak loads;
- System cycle time (how long it take to produce one part);
- Utilization of resource, labor, and machines;
- Bottlenecks and choke points;
- Queuing at work locations;
- Queuing and delays caused by material-handling devices and systems;
- WIP storage needs;
- Staffing requirements;
- Effectiveness of scheduling systems;
- Effectiveness of control systems.

More examples in various areas

Automobile simulator

A soldier tests out a heavy-wheeled-vehicle driver simulator.

An automobile simulator provides an opportunity to reproduce the characteristics of real vehicles in a virtual environment. It replicates the external factors and conditions with which a vehicle interacts enabling a driver to feel as if they are sitting in the cab of their own vehicle. Scenarios and events are replicated with sufficient reality to ensure that drivers become fully immersed in the experience rather than simply viewing it as an educational experience.

The simulator provides a constructive experience for the novice driver and enables more complex exercises to be undertaken by the more mature driver. For novice drivers, truck simulators provide an opportunity to begin their career by applying best practice. For mature drivers, simulation provides the ability to enhance good driving or to detect poor practice and to suggest the necessary steps for remedial action. For companies, it provides an opportunity to educate staff in the driving skills that achieve reduced maintenance costs, improved productivity and, most importantly, to ensure the safety of their actions in all possible situations.

Biomechanics simulators

A biomechanics simulator is used to analyze walking dynamics, study sports performance, simulate surgical procedures, analyze joint loads, design medical devices, and animate human and animal movement.

A neuromechanical simulator that combines biomechanical and biologically realistic neural network simulation. It allows the user to test hypotheses on the neural basis of behavior in a physically accurate 3-D virtual environment.

City and urban simulation

A city simulator can be a city-building game but can also be a tool used by urban planners to understand how cities are likely to evolve in response to various policy decisions. AnyLogic is an example of modern, large-scale urban simulators designed for use by urban planners. City simulators are generally agent-based simulations with explicit representations for land use and transportation. UrbanSim and LEAM are examples of large-scale urban simulation models that are used by metropolitan planning agencies and military bases for land use and transportation planning.

Classroom of the future

The "classroom of the future" will probably contain several kinds of simulators, in addition to textual and visual learning tools. This will allow students to enter the clinical years better prepared, and with a higher skill level. The advanced student or postgraduate will have a more concise and comprehensive method of retraining — or of incorporating new clinical procedures into their skill set — and regulatory bodies and medical institutions will find it easier to assess the proficiency and competency of individuals.

The classroom of the future will also form the basis of a clinical skills unit for continuing education of medical personnel; and in the same way that the use of periodic flight training assists airline pilots, this technology will assist practitioners throughout their career.

The simulator will be more than a "living" textbook, it will become an integral a part of the practice of medicine. The simulator environment will also provide a standard platform for curriculum development in institutions of medical education.

Communication satellite simulation

Modern satellite communications systems (SatCom) are often large and complex with many interacting parts and elements. In addition, the need for broadband connectivity on a moving vehicle has increased dramatically in the past few years for both commercial and military applications. To accurately predict and deliver high quality of service, satcom system designers have to factor in terrain as well as atmospheric and meteorological conditions in their planning. To deal with such complexity, system designers and operators increasingly turn towards computer models of their systems to simulate real world operational conditions and gain insights in to usability and requirements prior to final product sign-off. Modeling improves the understanding of the system by enabling the SatCom system designer or planner to simulate real world performance by injecting the models with multiple hypothetical atmospheric and environmental conditions.

Digital Lifecycle Simulation

Simulation of airflow over an engine

Simulation solutions are being increasingly integrated with CAx (CAD, CAM, CAE....) solutions and processes. The use of simulation throughout the product lifecycle, especially at the earlier concept and design stages, has the potential of providing substantial benefits. These benefits range from direct cost issues such as reduced prototyping and shorter time-to-market, to better performing products and higher margins. However, for some companies, simulation has not provided the expected benefits.

The research firm Aberdeen Group has found that nearly all best-in-class manufacturers use simulation early in the design process as compared to 3 or 4 laggards who do not.

The successful use of simulation, early in the lifecycle, has been largely driven by increased integration of simulation tools with the entire CAD, CAM and PLM solution-set. Simulation solutions can now function across the extended enterprise in a multi-CAD environment, and include solutions for managing simulation data and processes and ensuring that simulation results are made part of the product lifecycle history. The ability to use simulation across the entire lifecycle has been enhanced through improved user interfaces such as tailorable user interfaces and "wizards" which allow all appropriate PLM participants to take part in the simulation process.

Disaster preparedness and simulation training

Simulation training has become a method for preparing people for disasters. Simulations can replicate emergency situations and track how learners respond thanks to a lifelike experience. Disaster preparedness simulations can involve training on how to handle terrorism attacks, natural disasters, pandemic outbreaks, or other life-threatening emergencies.

One organization that has used simulation training for disaster preparedness is CADE (Center for Advancement of Distance Education). CADE[45] has used a video game to prepare emergency workers for multiple types of attacks. As reported by News-Medical.Net, "The video game is the first in a series of simulations to address bioterrorism, pandemic flu, smallpox and other disasters that emergency personnel must prepare for.[46] " Developed by a team from the University of Illinois at Chicago (UIC), the game allows learners to practice their emergency skills in a safe, controlled environment.

The Emergency Simulation Program (ESP) at the British Columbia Institute of Technology (BCIT), Vancouver, British Columbia, Canada is another example of an organization that uses simulation to train for emergency situations. ESP uses simulation to train on the following situations: forest fire fighting, oil or chemical spill response, earthquake response, law enforcement, municipal fire fighting, hazardous material handling, military training, and response to terrorist attack [47] One feature of the simulation system is the implementation of "Dynamic Run-Time Clock," which allows simulations to run a 'simulated' time frame, 'speeding up' or 'slowing down' time as desired"[47] Additionally, the system allows session recordings, picture-icon based navigation, file storage of individual simulations, multimedia components, and launch external applications.

At the University of Québec in Chicoutimi, a research team at the outdoor research and expertise laboratory (Laboratoire d'Expertise et de Recherche en Plein Air - LERPA) specializes in using wilderness backcountry accident simulations to verify emergency response coordination.

Instructionally, the benefits of emergency training through simulations are that learner performance can be tracked through the system. This allows the developer to make adjustments as necessary or alert the educator on topics that may require additional attention. Other advantages are that the learner can be guided or trained on how to respond appropriately before continuing to the next emergency segment—this is an aspect that may not be available in the live-environment. Some emergency training simulators also allows for immediate feedback, while other simulations may provide a summary and instruct the learner to engage in the learning topic again.

In a live-emergency situation, emergency responders do not have time to waste. Simulation-training in this environment provides an opportunity for learners to gather as much information as they can and practice their knowledge in a safe environment. They can make mistakes without risk of endangering lives and be given the opportunity to correct their errors to prepare for the real-life emergency.

Engineering, technology or process simulation

Simulation is an important feature in engineering systems or any system that involves many processes. For example in electrical engineering, delay lines may be used to simulate propagation delay and phase shift caused by an actual transmission line. Similarly, dummy loads may be used to simulate impedance without simulating propagation, and is used in situations where propagation is unwanted. A simulator may imitate only a few of the operations and functions of the unit it simulates. *Contrast with*: emulate.[48]

Most engineering simulations entail mathematical modeling and computer assisted investigation. There are many cases, however, where mathematical modeling is not reliable. Simulation of fluid dynamics problems often require both mathematical and physical simulations. In these cases the physical models require dynamic similitude. Physical and chemical simulations have also direct realistic uses,[49] rather than research uses; in chemical engineering, for example, process simulations are used to give the process parameters immediately used for operating chemical plants, such as oil refineries.

Economics simulation

In economics and especially macroeconomics, the effects of proposed policy actions, such as fiscal policy changes or monetary policy changes, are simulated to judge their desirability. A mathematical model of the economy, having been fitted to historical economic data, is used as a proxy for the actual economy; proposed values of government spending, taxation, open market operations, etc. are used as inputs to the simulation of the model, and various variables of interest such as the inflation rate, the unemployment rate, the balance of trade deficit, the government budget deficit, etc. are the outputs of the simulation. The simulated values of these variables of interest are compared for different proposed policy inputs to determine which set of outcomes is most desirable.

Equipment simulation

Due to the dangerous and expensive nature of training on heavy equipment, simulation has become a common solution across many industries. Types of simulated equipment include cranes, mining reclaimers and construction equipment, among many others. Often the simulation units will include pre-built scenarios by which to teach trainees, as well as the ability to customize new scenarios. Such equipment simulators are intended to create a safe and cost effective alternative to training on live equipment.[50]

Finance simulation

In finance, computer simulations are often used for scenario planning. Risk-adjusted net present value, for example, is computed from well-defined but not always known (or fixed) inputs. By imitating the performance of the project under evaluation, simulation can provide a distribution of NPV over a range of discount rates and other variables.

Simulations are frequently used in financial training to engage participants in experiencing various historical as well as fictional situations. There are stock market simulations, portfolio simulations, risk management simulations or models and forex simulations. Such simulations are typically based on stochastic asset models. Using these simulations in a training program allows for the application of theory into a something akin to real life. As with other industries, the use of simulations can be technology or case-study driven.

Flight simulation

Flight Simulation Training Devices (FSTD) are used to train pilots on the ground. In comparison to training in an actual aircraft, simulation based training allows for the training of maneuvers or situations that may be impractical (or even dangerous) to perform in the aircraft, while keeping the pilot and instructor in a relatively low-risk environment on the ground. For example, electrical system failures, instrument failures, hydraulic system failures, and even flight control failures can be simulated without risk to the pilots or an aircraft.

Instructors can also provide students with a higher concentration of training tasks in a given period of time than is usually possible in the aircraft. For example, conducting multiple instrument approaches in the actual aircraft may require significant time spent repositioning the aircraft, while in a simulation, as soon as one approach has been completed, the instructor can immediately preposition the simulated aircraft to an ideal (or less than ideal) location from which to begin the next approach.

Flight simulation also provides an economic advantage over training in an actual aircraft. Once fuel, maintenance, and insurance costs are taken into account, the operating costs of an FSTD are usually substantially lower than the operating costs of the simulated aircraft. For some large transport category airplanes, the operating costs may be several times lower for the FSTD than the actual aircraft.

Some people who use simulator software, especially flight simulator software, build their own simulator at home. Some people — to further the realism of their homemade simulator — buy used cards and racks that run the same software used by the original machine. While this involves solving the problem of matching hardware and software — and the problem that hundreds of cards plug into many different racks — many still find that solving these problems is well worthwhile. Some are so serious about realistic simulation that they will buy real aircraft parts, like complete nose sections of written-off aircraft, at aircraft boneyards. This permits people to simulate a hobby that they are unable to pursue in real life.

Marine simulators

Bearing resemblance to flight simulators, marine simulators train ships' personnel. The most common marine simulators include:

- Ship's bridge simulators
- Engine room simulators
- Cargo handling simulators
- Communication / GMDSS simulators
- ROV simulators

Simulators like these are mostly used within maritime colleges, training institutions and navies. They often consist of a replication of a ships' bridge, with operating console(s), and a number of screens on which the virtual surroundings are projected.

Military simulations

Military simulations, also known informally as war games, are models in which theories of warfare can be tested and refined without the need for actual hostilities. They exist in many different forms, with varying degrees of realism. In recent times, their scope has widened to include not only military but also political and social factors (for example, the NationLab series of strategic exercises in Latin America).[51] While many governments make use of simulation, both individually and collaboratively, little is known about the model's specifics outside professional circles.

Robotics simulators

A robotics simulator is used to create embedded applications for a specific (or not) robot without being dependent on the 'real' robot. In some cases, these applications can be transferred to the real robot (or rebuilt) without modifications. Robotics simulators allow reproducing situations that cannot be 'created' in the real world because of cost, time, or the 'uniqueness' of a resource. A simulator also allows fast robot prototyping. Many robot simulators feature physics engines to simulate a robot's dynamics.

Production simulation

Simulations of production systems is used mainly to examine the effect of improvements or investments in a production system. Most often this is done using a static spreadsheet with process times and transportation times. For more sophisticated simulations Discrete Event Simulation (DES) is used with the advantages to simulate dynamics in the production system. A production system is very much dynamic depending on variations in manufacturing processes, assembly times, machine set-ups, breaks, breakdowns and small stoppages.[52] There are lots of programs commonly used for discrete event simulation. They differ in usability and markets but do often share the same foundation. There is an academic project investigating the possibilities to use production simulation software for ecology labeling, named EcoProIT.

Sales process simulators

Simulations are useful in modeling the flow of transactions through business processes, such as in the field of sales process engineering, to study and improve the flow of customer orders through various stages of completion (say, from an initial proposal for providing goods/services through order acceptance and installation). Such simulations can help predict the impact of how improvements in methods might impact variability, cost, labor time, and the quantity of transactions at various stages in the process. A full-featured computerized process simulator can be used to depict such models, as can simpler educational demonstrations using spreadsheet software, pennies being transferred between cups based on the roll of a die, or dipping into a tub of colored beads with a scoop.[53]

Payment and securities settlement system simulations

Simulation techniques have also been applied to payment and securities settlement systems. Among the main users are central banks who are generally responsible for the oversight of market infrastructure and entitled to contribute to the smooth functioning of the payment systems.

Central banks have been using payment system simulations to evaluate things such as the adequacy or sufficiency of liquidity available (in the form of account balances and intraday credit limits) to participants (mainly banks) to allow efficient settlement of payments.[54] [55] The need for liquidity is also dependent on the availability and the type of netting procedures in the systems, thus some of the studies have a focus on system comparisons.[56]

Another application is to evaluate risks related to events such as communication network breakdowns or the inability of participants to send payments (e.g. in case of possible bank failure).[57] This kind of analysis falls under the concepts of Stress testing or scenario analysis.

A common way to conduct these simulations is to replicate the settlement logics of the real payment or securities settlement systems under analysis and then use real observed payment data. In case of system comparison or system development, naturally also the other settlement logics need to be implemented.

To perform stress testing and scenario analysis, the observed data needs to be altered, e.g. some payments delayed or removed. To analyze the levels of liquidity, initial liquidity levels are varried. System comparisons (benchmarking)or evaluations of new netting algorithms or rules are performed by running simulations with a fixed set of data and varying only the system setups.

Inference is usually done by comparing the benchmark simulation results to the results of altered simulation setups by comparing indicators such as unsettled transactions or settlement delays

Space shuttle countdown simulation

Firing Room 1 configured for space shuttle launches

Simulation is used at Kennedy Space Center (KSC) to train and certify Space Shuttle engineers during simulated launch countdown operations. The Space Shuttle engineering community participates in a launch countdown integrated simulation before each shuttle flight. This simulation is a virtual simulation where real people interact with simulated Space Shuttle vehicle and Ground Support Equipment (GSE) hardware. The Shuttle Final Countdown Phase Simulation, also known as S0044, involves countdown processes that integrate many of the Space Shuttle vehicle and GSE systems. Some of the Shuttle systems integrated in the simulation are the main propulsion system, main engines, solid rocket boosters, ground liquid hydrogen and liquid oxygen, external tank, flight controls, navigation, and avionics.[58] The high-level objectives of the Shuttle Final Countdown Phase Simulation are:

- To demonstrate Firing Room final countdown phase operations.
- To provide training for system engineers in recognizing, reporting and evaluating system problems in a time critical environment.
- To exercise the launch teams ability to evaluate, prioritize and respond to problems in an integrated manner within a time critical environment.

- To provide procedures to be used in performing failure/recovery testing of the operations performed in the final countdown phase.[59]

The Shuttle Final Countdown Phase Simulation takes place at the Kennedy Space Center Launch Control Center Firing Rooms. The firing room used during the simulation is the same control room where real launch countdown operations are executed. As a result, equipment used for real launch countdown operations is engaged. Command and control computers, application software, engineering plotting and trending tools, launch countdown procedure documents, launch commit criteria documents, hardware requirement documents, and any other items used by the engineering launch countdown teams during real launch countdown operations are used during the simulation. The Space Shuttle vehicle hardware and related GSE hardware is simulated by mathematical models (written in Shuttle Ground Operations Simulator (SGOS) modeling language [60]) that behave and react like real hardware. During the Shuttle Final Countdown Phase Simulation, engineers command and control hardware via real application software executing in the control consoles – just as if they were commanding real vehicle hardware. However, these real software applications do not interface with real Shuttle hardware during simulations. Instead, the applications interface with mathematical model representations of the vehicle and GSE hardware. Consequently, the simulations bypass sensitive and even dangerous mechanisms while providing engineering measurements detailing how the hardware would have reacted. Since these math models interact with the command and control application software, models and simulations are also used to debug and verify the functionality of application software.[61]

Satellite navigation simulators

The only true way to test GNSS receivers (commonly known as Sat-Nav's in the commercial world)is by using an RF Constellation Simulator. A receiver that may for example be used on an aircraft, can be tested under dynamic conditions without the need to take it on a real flight. The test conditions can be repeated exactly, and there is full control over all the test parameters. this is not possible in the 'real-world' using the actual signals. For testing receivers that will use the new Galileo (satellite navigation) there is no alternative, as the real signals do not yet exist.

Weather simulation

Predicting weather conditions by extrapolating/interpolating previous data is one of the real use of simulation. Most of the weather forecats use this information published by Weather buereaus. This kind of simulations help in predicting and forewarning about extreme weather conditions like the path of an active hurricane/cyclone. Numerical weather prediction for forecasting involves complicated numeric computer models to predict weather accurately by taking many parameters in to account.

Simulation and games

Strategy games — both traditional and modern — may be viewed as simulations of abstracted decision-making for the purpose of training military and political leaders (see History of Go for an example of such a tradition, or Kriegsspiel for a more recent example).

Many other video games are simulators of some kind. Such games can simulate various aspects of reality, from business, to government, to construction, to piloting vehicles (see above).

Historical usage

Historically, the word had negative connotations:

> ...for Distinction Sake, a Deceiving by Words, is commonly called a Lye, and a Deceiving by Actions, Gestures, or Behavior, is called Simulation...
>
> —Robert South, South, 1697, p.525

However, the connection between simulation and dissembling later faded out and is now only of linguistic interest.[62]

See also

- Dissimulation
- Emulator
- in silico
- Futures studies
- High-level emulation
- Lifelike experience
- List of discrete event simulation software
- List of computer simulation software
- Mathematical model
- Merger simulation
- Project Management Simulation
- Mining simulation
- Molecular dynamics
- Network Simulator
- Pharmacokinetics simulation
- Placebo
- Roleplay simulation
- Simulation language
- Similitude (model)
- Simulated reality
- Training Simulation
- Web based simulation

References

[1] J. Banks, J. Carson, B. Nelson, D. Nicol (2001). *Discrete-Event System Simulation*. Prentice Hall. p. 3. ISBN 0-13-088702-1.

[2] In the words of the Simulation article (http://www.modelbenders.com/encyclopedia/encyclopedia.html) in Encyclopedia of Computer Science, "designing a model of a real or imagined system and conducting experiments with that model".

[3] Sokolowski, J.A., Banks, C.M. (2009). *Principles of Modeling and Simulation*. Hoboken, NJ: Wiley. p. 6. ISBN 978-0-470-28943-3.

[4] For example in computer graphics SIGGRAPH 2007 | For Attendees | Papers (http://www.siggraph.org/s2007/attendees/papers/12.html) Doc:Tutorials/Physics/BSoD - BlenderWiki (http://wiki.blender.org/index.php/BSoD/Physical_Simulation).

[5] Thales defines synthetic environment as "the counterpart to simulated models of sensors, platforms and other active objects" for "the simulation of the external factors that affect them" (http://www.thalesresearch.com/Default.aspx?tabid=181) while other vendors use the term for more visual, virtual reality-style simulators (http://www.cae.com/www2004/Products_and_Services/Civil_Simulation_and_Training/Simulation_Equipment/Visual_Solutions/Synthetic_Environments/index.shtml).

[6] For a popular research project in the field of biochemistry where "computer simulation is particularly well suited to address these questions" Folding@home - Main (http://folding.stanford.edu/Pande/Main), see Folding@Home.

[7] For an academic take on a training simulator, see e.g. Towards Building an Interactive, Scenario-based Training Simulator (http://gel.msu.edu/magerko/papers/11TH-CGF-058.pdf), for medical application Medical Simulation Training Benefits (http://www.immersion.com/medical/benefits1.php) as presented by a simulator vendor and for military practice A civilian's guide to US defense and security assistance to Latin America and the Caribbean (http://ciponline.org/facts/exe.htm) published by Center for International Policy.

[8] http://www.simschool.org

[9] Classification used by the Defense Modeling and Simulation Office.

[10] "High Versus Low Fidelity Simulations: Does the Type of Format Affect Candidates' Performance or Perceptions?" (http://www.ipmaac.org/conf/03/havighurst.pdf)

[11] For example All India management association (http://www.aima-ind.org/) maintains that playing to win, participants "imbibe new forms of competitive behavior that are ideal for today's highly chaotic business conditions" (http://www.aima-ind.org/management_games.asp) and IBM claims that "the skills honed playing massive multiplayer dragon-slaying games like World of Warcraft can be useful when managing modern multinationals".

[12] "Reacting to the Past Home Page" (http://www.barnard.columbia.edu/reacting/)

[13] "Carana," at 'PaxSims' blog, 27 January 2009 (http://paxsims.wordpress.com/2009/01/27/carana/)

[14] Sherman, W.R., Craig, A.B. (2003). *Understanding Virtual Reality*. San Francisco, CA: Morgan Kaufmann. ISBN 978-1-55860-353-0.

[15] Leeb, R., Lee, F., Keinrath, C., Schere, R., Bischof, H., Pfurtscheller, G. (2007). "Brain-Computer Communication: Motivation, Aim, and Impact of Exploring a Virtual Apartment". *IEEE Transactions on Neural Systems and Rehabilitation Engineering* **15** (4): 473–481. doi:10.1109/TNSRE.2007.906956.

[16] Zahraee, A.H., Szewczyk, J., Paik, J.K., Guillaume, M. (2010). Robotic hand-held surgical device: evaluation of end-effector's kinematics and development of proof-of-concept prototypes. Proceedings of the 13th International Conference on Medical Image Computing and Computer Assisted Intervention, Beijing, China.

[17] Ahmed K, Keeling AN, Fakhry M, Ashrafian H, Aggarwal R, Naughton PA, Darzi A, Cheshire N, *et al.* (January 2010). "Role of Virtual Reality Simulation in Teaching and Assessing Technical Skills in Endovascular Intervention". *J Vasc Interv Radiol* **21**.

[18] Narayan, Roger; Kumta, Prashant; Sfeir, Charles; Lee, Dong-Hyun; Choi, Daiwon; Olton, Dana (October 2004). "Nanostructured ceramics in medical devices: Applications and prospects" (http://www.ingentaconnect.com/content/tms/jom/2004/00000056/00000010/art00011). *JOM* **56** (10): 38–43. Bibcode 2004JOM....56j..38N. doi:10.1007/s11837-004-0289-x. PMID 11196953. .

[19] Couvreur P, Vauthier C (July 2006). "Nanotechnology: intelligent design to treat complex disease". *Pharm. Res.* **23** (7): 1417–50. doi:10.1007/s11095-006-0284-8. PMID 16779701.

[20] Hede S, Huilgol N (2006). ""Nano": the new nemesis of cancer" (http://www.cancerjournal.net/article.asp?issn=0973-1482;year=2006;volume=2;issue=4;spage=186;epage=195;aulast=Hede). *J Cancer Res Ther* **2** (4): 186–95. doi:10.4103/0973-1482.29829. PMID 17998702. .

[21] Leary SP, Liu CY, Apuzzo ML (June 2006). "Toward the emergence of nanoneurosurgery: part III—nanomedicine: targeted nanotherapy, nanosurgery, and progress toward the realization of nanoneurosurgery" (http://meta.wkhealth.com/pt/pt-core/template-journal/lwwgateway/media/landingpage.htm?issn=0148-396X&volume=58&issue=6&spage=1009). *Neurosurgery* **58** (6): 1009–26; discussion 1009–26. doi:10.1227/01.NEU.0000217016.79256.16. PMID 16723880. .

[22] http://vam.anest.ufl.edu/wip.html

[23] Nishisaki A, Keren R, Nadkarni V (June 2007). "Does simulation improve patient safety? Self-efficacy, competence, operational performance, and patient safety" (http://linkinghub.elsevier.com/retrieve/pii/S1932-2275(07)00025-0). *Anesthesiol Clin* **25** (2): 225–36. doi:10.1016/j.anclin.2007.03.009. PMID 17574187. .

[24] Meller, G. (1997). "A Typology of Simulators for Medical Education" (http://www.medsim.com/profile/article1.html). *Journal of Digital Imaging*. .

[25] Cooper Jeffery B, Taqueti VR (2008-12). "A brief history of the development of mannequin simulators for clinical education and training" (http://pmj.bmj.com/content/84/997/563.long). *Postgrad Med J.* **84** (997): 563–570. doi:10.1136/qshc.2004.009886. PMID 19103813. . Retrieved 2011-05-24.

[26] Murphy D, Challacombe B, Nedas T, Elhage O, Althoefer K, Seneviratne L, Dasgupta P. (May 2007). "[Equipment and technology in robotics]" (in Spanish; Castilian). *Arch. Esp. Urol.* **60** (4): 349–55. PMID 17626526.

[27] "In Cybertherapy, Avatars Assist With Healing" (http://www.nytimes.com/2010/11/23/science/23avatar.html?_r=1&ref=science). *New York Times*. 2010-11-22. . Retrieved 2010-11-23.

[28] Dagger, Jacob (May–June 2008). *Update: "The New Game Theory"* (http://www.dukemagazine.duke.edu/dukemag/issues/050608/depupd.html). **94**. Duke Magazine. . Retrieved 2011-02-08.

[29] Steinberg, Scott (2011-01-31). "How video games can make you smarter" (http://articles.cnn.com/2011-01-31/tech/video.games.smarter.steinberg_1_video-games-interactive-simulations-digital-world?_s=PM:TECH). *Cable News Network (CNN Tech)*. . Retrieved 2011-02-08.

[30] Vlaovic PD, Sargent ER, Boker JR, *et al.* (2008). "Immediate impact of an intensive one-week laparoscopy training program on laparoscopic skills among postgraduate urologists" (http://openurl.ingenta.com/content/nlm?genre=article&issn=1086-8089&volume=12&issue=1&spage=1&aulast=Vlaovic). *JSLS* **12** (1): 1–8. PMC 3016039. PMID 18402731. .

[31] Leung J, Foster E (April 2008). "How do we ensure that trainees learn to perform biliary sphincterotomy safely, appropriately, and effectively?" (http://www.current-reports.com/article_frame.cfm?PubID=GR10-2-2-03&Type=Abstract). *Curr Gastroenterol Rep* **10** (2): 163–8. doi:10.1007/s11894-008-0038-3. PMID 18462603. .

[32] http://www.pong-story.com/intro.htm Archived (http://www.webcitation.org/5seElSgBC) at WebCite

[33] http://homepages.vvm.com/~jhunt/compupedia/History%20of%20Computers/history_of_computers_1980.htm

[34] "Video Game Console Timeline - Video Game History - Xbox 360 - TIME Magazine" (http://www.time.com/time/covers/1101050523/console_timeline/). *Time*. 2005-05-23. . Retrieved 2010-05-23.

[35] http://design.osu.edu/carlson/history/tron.html

[36] http://www.beanblossom.in.us/larryy/cgi.html

[37] http://www.starksravings.com/linktrainer/linktrainer.htm

[38] http://www.trudang.com/simulatr/simulatr.html

[39] http://open-site.org/Games/Video_Games/Simulation

[40] http://www.ibisworld.com/industry/retail.aspx?indid=2003&chid=1

[41] http://www.sciencedaily.com/articles/c/computer-generated_imagery.htm

[42] http://www.awn.com/mag/issue4.02/4.02pages/kenyonspiderman.php3

[43] Benedettini, O., Tjahjono, B. (2008). "Towards an improved tool to facilitate simulation modeling of complex manufacturing systems". *International Journal of Advanced Manufacturing Technology* **43** (1/2): 191–9. doi:10.1007/s00170-008-1686-z.

[44] Banks, J., Carson J., Nelson B.L., Nicol, D. (2005). *Discrete-event system simulation* (4th ed.). Upper Saddle River, NJ: Pearson Prentice Hall. ISBN 978-0-13-088702-3.

[45] CADE- http://www.uic.edu/sph/cade/

[46] News-Medical.Net article- http://www.news-medical.net/news/2005/10/27/14106.aspx

[47] http://www.straylightmm.com/

[48] Federal Standard 1037C

[49] D. Passeri et al. (May 2009). "Analysis of 3D stacked fully functional CMOS Active Pixel Sensor detectors" (http://meroli.web.cern.ch/meroli/Analysisof3Dstacked.html). *Journal of Instrumentation* **4** (4): 4009. doi:10.1088/1748-0221/4/04/P04009. .

[50] http://www.globalsim.com

[51] The Economist provides a current (as of 2012) survey of public projects attempting to simulate some theories in "The science of civil war: What makes heroic strife" (http://www.economist.com/node/21553006).

[52] Ulf, Eriksson (2005). *Diffusion of Discrete Event Simulation in Swedish Industry*. Gothenburg: Doktorsavhandlingar vid Chalmers tekniska högskola. ISBN 91-7291-577-3.

[53] Paul H. Selden (1997). ***Sales Process Engineering: A Personal Workshop***. Milwaukee, WI: ASQ Quality Press. ISBN 978-0-87389-418-0.

[54] Leinonen (ed.): Simulation studies of liquidity needs, risks and efficiency in payment networks (Bank of Finland Studies E:39/2007) Simulation publications (http://pss.bof.fi/Pages/Publications.aspx)

[55] Neville Arjani: Examining the Trade-Off between Settlement Delay and Intraday Liquidity in Canada's LVTS: A Simulation Approach (Working Paper 2006-20, Bank of Canada) Simulation publications (http://pss.bof.fi/Pages/Publications.aspx)

[56] Johnson, K. - McAndrews, J. - Soramäki, K. 'Economizing on Liquidity with Deferred Settlement Mechanisms' (Reserve Bank of New York Economic Policy Review, December 2004)

[57] H. Leinonen (ed.): Simulation analyses and stress testing of payment networks (Bank of Finland Studies E:42/2009) Simulation publications (http://pss.bof.fi/Pages/Publications.aspx)

[58] Sikora, E.A. (2010, July 27). Space Shuttle Main Propulsion System expert, John F. Kennedy Space Center. Interview.

[59] Shuttle Final Countdown Phase Simulation. National Aeronautics and Space Administration KSC Document # RTOMI S0044, Revision AF05, 2009.

[60] Shuttle Ground Operations Simulator (SGOS) Summary Description Manual. National Aeronautics and Space Administration KSC Document # KSC-LPS-SGOS-1000, Revision 3 CHG-A, 1995.

[61] Math Model Main Propulsion System (MPS) Requirements Document, National Aeronautics and Space Administration KSC Document # KSCL-1100-0522, Revision 9, June 2009.

[62] South, in the passage quoted, was speaking of the differences between a falsehood and an honestly mistaken statement; the difference being that in order for the statement to be a lie the truth must be known, and the opposite of the truth must have been knowingly uttered. And, from this, to the extent to which a **lie** involves deceptive *words*, a **simulation** involves deceptive *actions*, deceptive *gestures*, or deceptive *behavior*. Thus, it would seem, if a simulation is **false**, then the truth must be known (in order for *something other than the truth* to be presented in its stead); and, for the **simulation** to *simulate*. Because, otherwise, one would not know what to offer up in simulation. Bacon's essay Of Simulation and Dissimulation (http://www.authorama.com/essays-of-francis-bacon-7.html) expresses somewhat similar views; it is also significant that Samuel Johnson thought so highly of South's definition, that he used it in the entry for simulation in his *Dictionary of the English Language*.

Further reading

- C. Aldrich (2003). *Learning by Doing : A Comprehensive Guide to Simulations, Computer Games, and Pedagogy in e-Learning and Other Educational Experiences*. San Francisco: Pfeifer — John Wiley & Sons. ISBN 978-0-7879-7735-1.
- C. Aldrich (2004). *Simulations and the future of learning: an innovative (and perhaps revolutionary) approach to e-learning*. San Francisco: Pfeifer — John Wiley & Sons. ISBN 978-0-7879-6962-2.
- Steve Cohen (2006). *Virtual Decisions*. Mahwah, NJ: Lawrence Erlbaum Associates. ISBN 978-0-8058-4994-3.
- R. Frigg, S. Hartmann (2007). "Models in Science" (http://plato.stanford.edu/entries/models-science/). *Stanford Encyclopedia of Philosophy*.
- S. Hartmann (1996). "The World as a Process: Simulations in the Natural and Social Sciences" (http://philsci-archive.pitt.edu/archive/00002412/). In R. Hegselmann, *et al.*. *Modelling and Simulation in the Social Sciences from the Philosophy of Science Point of View*. Theory and Decision Library. Dordrecht: Kluwer. pp. 77–100.
- J.P. Hertel (2002). *Using Simulations to Promote Learning in Higher Education*. Sterling, Virginia: Stylus. ISBN 978-1-57922-052-5.
- P. Humphreys (2004). *Extending Ourselves: Computational Science, Empiricism, and Scientific Method*. Oxford: Oxford University Press. ISBN 978-0-19-515870-0.
- F. Percival, S. Lodge, D. Saunders (1993). *The Simulation and Gaming Yearbook: Developing Transferable Skills in Education and Training*. London: Kogan Page.

- James J. Nutaro, *Building Software for Simulation: Theory and Algorithms, with Applications in C++*. Wiley, 2010.
- D. Saunders, ed. (2000). *The International Simulation and Gaming Research Yearbook*. London: Kogan Page.
- Roger D. Smith: Simulation Article (http://www.modelbenders.com/encyclopedia/encyclopedia.html), Encyclopedia of Computer Science, Nature Publishing Group, ISBN 978-0-333-77879-1.
- Roger D. Smith: "Simulation: The Engine Behind the Virtual World" (http://www.modelbenders.com/Bookshop/techpapers.html), eMatter, December, 1999.
- R. South (1688). "A Sermon Delivered at Christ-Church, Oxon., Before the University, Octob. 14. 1688: Prov. XII.22 Lying Lips are abomination to the Lord", pp. 519–657 in South, R., *Twelve Sermons Preached Upon Several Occasions (Second Edition), Volume I*, Printed by S.D. for Thomas Bennet, (London), 1697.
- Gabriel A. Wainer (2009) Discrete-Event Modeling and Simulation: a practitioner's approach (http://cell-devs.sce.carleton.ca/mediawiki/index.php/Main_Page). CRC Press, 2009.
- Eric Winsberg (1999) Sanctioning Models: The epistemology of simulation (http://www.cas.usf.edu/~ewinsb/SiC_Eric_Winsberg.pdf), in Sismondo, Sergio and Snait Gissis (eds.) (1999), Modeling and Simulation. Special Issue of Science in Context 12.
- Eric Winsberg (2001). "Simulations, Models and Theories: Complex Physical Systems and their Representations". *Philosophy of Science* **68**: 442–454.
- Eric Winsberg (2003). "Simulated Experiments: Methodology for a Virtual World" (http://www.cas.usf.edu/~ewinsb/methodology.pdf) (PDF). *Philosophy of Science* **70**: 105–125. doi:10.1086/367872.
- Joseph Wolfe, David Crookall (1998). "Developing a scientific knowledge of simulation/gaming" (http://sag.sagepub.com/cgi/reprint/29/1/7). *Simulation & Gaming: an International Journal of Theory, Design and Research* **29** (1): 7–19.
- Ellen K. Levy (2004). "Synthetic Lighting: Complex Simulations of Nature". *Photography Quarterly* (88): 5–9.
- Paul Humphreys and Cyrille Imbert, ed. (2012). *Models, Simulations and Representations*. London and New York: Routledge. ISBN 978-0-203-80841-2 (ebk).

External links

- **Bibliographies** containing more references (http://www.unice.fr/sg/resources/bibliographies.htm) to be found on the website of the journal *Simulation & Gaming* (http://www.unice.fr/sg/).
- **What are simulations good for?** (http://www.learningsutras.com/why-are-simulations-good) A Focalworks Q&A.

Photography

Photography is the art, science and practice of creating durable images by recording light or other electromagnetic radiation, either electronically by means of an image sensor or chemically by means of a light-sensitive material such as photographic film.[1] Typically, a lens is used to focus the light reflected or emitted from objects into a real image on the light-sensitive surface inside a camera during a timed exposure. The result in an electronic image sensor is an electrical charge at each pixel, which is electronically processed and stored in a digital image file for subsequent display or processing. The result in a photographic emulsion is an invisible latent image, which is later chemically developed into a visible image, either negative or positive depending on the purpose of the photographic material and the method of processing. A negative image on film is traditionally used to photographically create a positive image on a paper base, known as a print, either by using an enlarger or by contact printing.

Photography has many uses for business, science, manufacturing (e.g. photolithography), art, and recreational purposes.

Etymology

Lens and mounting of a large-format camera

As far as can be ascertained, it was Sir John Herschel in a lecture before the Royal Society of London, on March 14, 1839 who made the word "photography" known to the world. But in an article published on February 25 of the same year in a German newspaper called the *Vossische Zeitung*, Johann von Maedler, a Berlin astronomer, had used the word photography already.[2] The word photography derives from the Greek φωτός (*phōtos*), genitive of φῶς (*phōs*), "light"[3] and γραφή (*graphé*) "representation by means of lines" or "drawing",[4] together meaning "drawing with light".[5]

Function

The Contax S of 1949 – the first pentaprism SLR

The camera is the image-forming device, and photographic film or a silicon electronic image sensor is the sensing medium. The respective recording medium can be the film itself, or a digital electronic or magnetic memory.[6]

Photographers control the camera and lens to "expose" the light recording material (such as film) to the required amount of light to form a "latent image" (on film) or "raw file" (in digital cameras) which, after appropriate processing, is converted to a usable image. Digital cameras use an electronic image sensor based on light-sensitive electronics such as charge-coupled device (CCD) or complementary metal-oxide-semiconductor (CMOS) technology. The resulting digital image is stored electronically, but can be reproduced on paper or film.

The camera (or 'camera obscura') is a dark room or chamber from which, as far as possible, all light is excluded except the light that forms the image. The subject being photographed, however, must be illuminated. Cameras can range from small to very large, a whole room that is kept dark while the object to be photographed is in another room where it is properly illuminated. This was common for reproduction photography of flat copy when large film negatives were used. A general principle known from the birth of photography is that the smaller the camera, the brighter the image. This meant that as soon as photographic materials became sensitive enough (*fast* enough) to take candid or what were called genre pictures, small detective cameras were used, some of them disguised as a tie pin that was really a lens, as a piece of luggage or even a pocket watch (the *Ticka* camera).

The black body Nikon F2AS, shown here, was very popular amongst photojournalists in the 1970's.

The discovery of the 'camera obscura' that provides an image of a scene is very old, dating back to ancient China. Leonardo da Vinci mentions natural camera obscuras that are formed by dark caves on the edge of a sunlit valley. A hole in the cave wall will act as a pinhole camera and project a laterally reversed, upside down image on a piece of paper. So the invention of photography was really concerned with finding a means to fix and retain the image in the camera obscura. This in fact occurred first using the reproduction of images without a camera when Josiah Wedgewood, from the famous family of potters, obtained copies of paintings on leather using silver salts. As he had no way of fixing them, that is to say to stabilize the image by washing out the non-exposed silver salts, they turned completely black in the light and had to be kept in a dark room for viewing.

Nikon F of 1959 — an early example of 35 mm film system camera

Renaissance painters used the camera obscura which, in fact, gives the optical rendering in color that dominates Western Art. The camera obscura literally means "dark chamber" in Latin. It is a box with a hole in it which allows light to go through and create an image onto the piece of paper.

Late production Minox B camera with later style "honeycomb" selenium light meter

The movie camera is a type of photographic camera which takes a rapid sequence of photographs on strips of film. In contrast to a still camera, which captures a single snapshot at a time, the movie camera takes a series of images, each called a "frame". This is accomplished through an intermittent mechanism. The frames are later played back in a movie projector at a specific speed, called the "frame rate" (number of frames per second). While viewing, a person's eyes and brain merge the separate pictures together to create the illusion of motion.[7]

A modern digital camera

In all but certain specialized cameras, the process of obtaining a usable exposure must involve the use, manually or automatically, of a few controls to ensure the photograph is clear, sharp and well illuminated. The controls usually include but are not limited to the following:

Control	Description
Focus	The position of a viewed object or the adjustment of an optical device necessary to produce a clear image: in focus; out of focus.[8]
Aperture	Adjustment of the lens opening, measured as f-number, which controls the amount of light passing through the lens. Aperture also has an effect on depth of field and diffraction – the higher the f-number, the smaller the opening, the less light, the greater the depth of field, and the more the diffraction blur. The focal length divided by the f-number gives the effective aperture diameter.
Shutter speed	Adjustment of the speed (often expressed either as fractions of seconds or as an angle, with mechanical shutters) of the shutter to control the amount of time during which the imaging medium is exposed to light for each exposure. Shutter speed may be used to control the amount of light striking the image plane; 'faster' shutter speeds (that is, those of shorter duration) decrease both the amount of light and the amount of image blurring from motion of the subject and/or camera.
White balance	On digital cameras, electronic compensation for the color temperature associated with a given set of lighting conditions, ensuring that white light is registered as such on the imaging chip and therefore that the colors in the frame will appear natural. On mechanical, film-based cameras, this function is served by the operator's choice of film stock or with color correction filters. In addition to using white balance to register natural coloration of the image, photographers may employ white balance to aesthetic end, for example white balancing to a blue object in order to obtain a warm color temperature.
Metering	Measurement of exposure so that highlights and shadows are exposed according to the photographer's wishes. Many modern cameras meter and set exposure automatically. Before automatic exposure, correct exposure was accomplished with the use of a separate light metering device or by the photographer's knowledge and experience of gauging correct settings. To translate the amount of light into a usable aperture and shutter speed, the meter needs to adjust for the sensitivity of the film or sensor to light. This is done by setting the "film speed" or ISO sensitivity into the meter.
ISO speed	Traditionally used to "tell the camera" the film speed of the selected film on film cameras, ISO speeds are employed on modern digital cameras as an indication of the system's *gain* from light to numerical output and to control the automatic exposure system. The higher the ISO number the greater the film sensitivity to light, whereas with a lower ISO number, the film is less sensitive to light. A correct combination of ISO speed, aperture, and shutter speed leads to an image that is neither too dark nor too light, hence it is 'correctly exposed', indicated by a centered meter.
Autofocus point	On some cameras, the selection of a point in the imaging frame upon which the auto-focus system will attempt to focus. Many Single-lens reflex cameras (SLR) feature multiple auto-focus points in the viewfinder.

Many other elements of the imaging device itself may have a pronounced effect on the quality and/or aesthetic effect of a given photograph; among them are:

- **Focal length** and **type of lens** (normal, long focus, wide angle, telephoto, macro, fisheye, or zoom)
- **Filters** placed between the subject and the light recording material, either in front of or behind the lens
- Inherent **sensitivity** of the medium to light intensity and color/wavelengths.
- The nature of the light **recording material**, for example its resolution as measured in pixels or grains of silver halide.

Exposure and rendering

Camera controls are interrelated. The total amount of light reaching the film plane (the 'exposure') changes with the duration of exposure, aperture of the lens, and on the effective focal length of the lens (which in variable focal length lenses, can force a change in aperture as the lens is zoomed). Changing any of these controls can alter the exposure. Many cameras may be set to adjust most or all of these controls automatically. This automatic functionality is useful for occasional photographers in many situations.

The duration of an exposure is referred to as shutter speed, often even in cameras that do not have a physical shutter, and is typically measured in fractions of a second. It is quite possible to have exposures one of several seconds, usually for still-life subjects, and for night scenes exposure times can be several hours.

The effective aperture is expressed by an f-number or f-stop (derived from focal ratio), which is proportional to the ratio of the focal length to the diameter of the aperture. Longer lenses will pass less light even though the diameter of the aperture is the same due to the greater distance the light has to travel: shorter lenses (a shorter focal length) will be brighter with the same size of aperture.

The smaller the f/number, the larger the effective aperture. The present system of f/numbers to give the effective aperture of a lens was standardized by an international convention. There were earlier, different series of numbers in older cameras.

If the f-number is decreased by a factor of , the aperture diameter is increased by the same factor, and its area is increased by a factor of 2. The f-stops that might be found on a typical lens include 2.8, 4, 5.6, 8, 11, 16, 22, 32, where going up "one stop" (using lower f-stop numbers) doubles the amount of light reaching the film, and stopping down one stop halves the amount of light.

Image capture can be achieved through various combinations of shutter speed, aperture, and film or sensor speed. Different (but related) settings of aperture and shutter speed enable photographs to be taken under various conditions of film or sensor speed, lighting and motion of subjects and/or camera, and desired depth of field. A slower speed film will exhibit less "grain", and a slower speed setting on an electronic sensor will exhibit less "noise", while higher film and sensor speeds allow for a faster shutter speed, which reduces motion blur or allows the use of a smaller aperture to increase the depth of field. For example, a wider aperture is used for lower light and a lower aperture for more light. If a subject is in motion, then a high shutter speed may be needed. A tripod can also be helpful in that it enables a slower shutter speed to be used.

For example, f/8 at 8 ms (1/125 of a second) and f/5.6 at 4 ms (1/250 of a second) yield the same amount of light. The chosen combination has an impact on the final result. The aperture and focal length of the lens determine the depth of field, which refers to the range of distances from the lens that will be in focus. A longer lens or a wider aperture will result in "shallow" depth of field (i.e. only a small plane of the image will be in sharp focus). This is often useful for isolating subjects from backgrounds as in individual portraits or macro photography. Conversely, a shorter lens, or a smaller aperture, will result in more of the image being in focus. This is generally more desirable when photographing landscapes or groups of people. With very small apertures, such as pinholes, a wide range of distance can be brought into focus, but sharpness is severely degraded by diffraction with such small apertures. Generally, the highest degree of "sharpness" is achieved at an aperture near the middle of a lens's range (for example, f/8 for a lens with available apertures of f/2.8 to f/16). However, as lens technology improves, lenses are becoming capable of making increasingly sharp images at wider apertures.

Image capture is only part of the image forming process. Regardless of material, some process must be employed to render the latent image captured by the camera into a viewable image. With slide film, the developed film is just mounted for projection. Print film requires the developed film negative to be printed onto photographic paper or transparency. Digital images may be uploaded to an image server (e.g., a photo-sharing web site), viewed on a television, or transferred to a computer or digital photo frame.

Prior to the rendering of a viewable image, modifications can be made using several controls. Many of these controls are similar to controls during image capture, while some are exclusive to the rendering process. Most printing controls have equivalent digital concepts, but some create different effects. For example, dodging and burning controls are different between digital and film processes. Other printing modifications include:

A photographer using a tripod for greater stability during long exposure.

- Chemicals and process used during film development
- Duration of print exposure – equivalent to shutter speed
- Printing aperture – equivalent to aperture, but has no effect on depth of field

- Contrast – changing the visual properties of objects in an image to make them distinguishable from other objects and the background
- Dodging – reduces exposure of certain print areas, resulting in lighter areas
- Burning in – increases exposure of certain areas, resulting in darker areas
- Paper texture – glossy, matte, etc.
- Paper type – resin-coated (RC) or fiber-based (FB)
- Paper size
- Toners – used to add warm or cold tones to black-and-white prints

Uses

Photography gained the interest of many scientists and artists from its inception. Scientists have used photography to record and study movements, such as Eadweard Muybridge's study of human and animal locomotion in 1887. Artists are equally interested by these aspects but also try to explore avenues other than the photo-mechanical representation of reality, such as the pictorialist movement. Military, police, and security forces use photography for surveillance, recognition and data storage. Photography is used by amateurs to preserve memories of favorite times, to capture special moments, to tell stories, to send messages, and as a source of entertainment.

Dark Room

History

Photography is the result of combining several technical discoveries. Long before the first photographs were made, Chinese philosopher Mo Di and Greek mathematicians Aristotle and Euclid described a pinhole camera in the 5th and 4th centuries BC.[10] [11] In the 6th century AD, Byzantine mathematician Anthemius of Tralles used a type of camera obscura in his experiments,[12] Ibn al-Haytham (Alhazen) (965–1040) studied the camera obscura and pinhole camera,[11] [13] Albertus Magnus (1193–1280) discovered silver nitrate,[14] and Georges Fabricius (1516–71) discovered silver chloride.[15] Daniele Barbaro described a diaphragm in 1568.[16] Wilhelm Homberg described how light darkened some chemicals (photochemical effect) in 1694.[17] The fiction book Giphantie, published in 1760, by French author Tiphaigne de la Roche, described what can be interpreted as photography.[16]

Earliest known surviving heliographic engraving, 1825, printed from a metal plate made by Joseph Nicéphore Niépce with his "heliographic process".[9] The plate was exposed under an ordinary engraving and copied it by photographic means. This was a step towards the first permanent photograph from nature taken with a camera obscura, in 1826.

Invented in the first decades of the 19th century, photography (by way of the camera) seemed able to capture more detail and information than traditional mediums, such as painting and sculpting.[18] Photography as a usable process goes back to the 1820s with the development of chemical photography. The first permanent photoetching was an image produced in 1822[9] by the French inventor Nicéphore

Niépce, but it was destroyed by a later attempt to duplicate it.[9] Niépce was successful again in 1825. He made the first permanent photograph from nature (his *View from the Window at Le Gras*) with a camera obscura in 1826.[19] However, because his photographs took so long to expose (eight hours), he sought to find a new process. Working in conjunction with Louis Daguerre, they experimented with silver compounds based on a Johann Heinrich Schultz discovery in 1816 that a silver and chalk mixture darkens when exposed to light. Niépce died in 1833, but Daguerre continued the work, eventually culminating with the development of the daguerreotype in 1837. Daguerre took the first ever photo of a person in 1838 when, while taking a daguerreotype of a Paris street, a pedestrian stopped for a shoe shine, long enough to be captured by the long exposure (several minutes). Eventually, France agreed to pay Daguerre a pension for his formula, in exchange for his promise to announce his discovery to the world as the gift of France, which he did in 1839.

Meanwhile, Hercules Florence had already created a very similar process in 1832, naming it *Photographie*, and English inventor William Fox Talbot had earlier discovered another means to fix a silver process image but had kept it secret. After reading about Daguerre's invention, Talbot refined his process so that portraits were made readily available to the masses. By 1840, Talbot had invented the calotype process, which creates negative images.[20] Talbot's famous 1835 print of the Oriel window in Lacock Abbey is the oldest known negative in existence.[21] [22] John Herschel made many contributions to the new methods. He invented the cyanotype process, now familiar as the "blueprint". He was the first to use the terms "photography", "negative" and "positive". He discovered sodium thiosulphate solution to be a solvent of silver halides in 1819, and informed Talbot and Daguerre of his discovery in 1839 that it could be used to "fix" pictures and make them permanent. He made the first glass negative in late 1839.

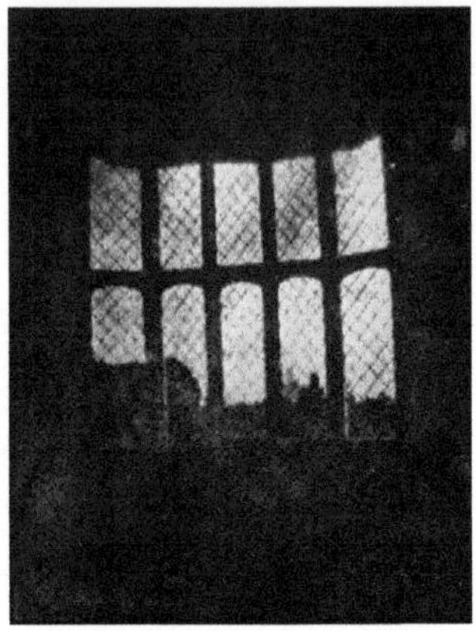

A latticed window in Lacock Abbey, England, photographed by William Fox Talbot in 1835. Shown here in positive form, this is the oldest known extant photographic negative made in a camera.

Mid-19th-century "Brady stand" photo model's armrest table, meant to keep portrait models more still during long exposure times (studio equipment nicknamed after the famed US photographer, Mathew Brady).

In March 1851, Frederick Scott Archer published his findings in "The Chemist" on the wet plate collodion process. This became the most widely used process between 1852 and the late 1860s when the dry plate was introduced. There are three subsets to the collodion process; the Ambrotype (positive image on glass), the Ferrotype or Tintype (positive image on metal) and the negative which was printed on albumen or salt paper.

Many advances in photographic glass plates and printing were made in through the 19th century. In 1884, George Eastman developed the technology of film to replace photographic plates, leading to the technology used by film cameras today.

In 1908 Gabriel Lippmann won the Nobel Laureate in Physics for his method of reproducing colors photographically based on the phenomenon of interference, also known as the Lippmann plate.

Processes

Black-and-white

All photography was originally monochrome, or *black-and-white*. Even after color film was readily available, black-and-white photography continued to dominate for decades, due to its lower cost and its "classic" photographic look. It is important to note that some monochromatic pictures are not always pure blacks and whites, but also contain other hues depending on the process. The cyanotype process produces an image of blue and white for example. The albumen process, first used more than 150 years ago, produces brown tones.

Many photographers continue to produce some monochrome images, often because of the established archival permanence of well processed silver halide based materials.

Some full color digital images are processed using a variety of techniques to create black and whites, and some manufacturers produce digital cameras that exclusively shoot monochrome.

Color

Color photography was explored beginning in the mid-19th century. Early experiments in color required extremely long exposures (hours or days for camera images) and could not "fix" the photograph to prevent the color from quickly fading when exposed to white light.

Early color photograph taken by Prokudin-Gorskii (1915).

The first permanent color photograph was taken in 1861 using the three-color-separation principle first published by physicist James Clerk Maxwell in 1855. Maxwell's idea was to take three separate black-and-white photographs through red, green and blue filters. This provides the photographer with the three basic channels required to recreate a color image. Transparent prints of the images could be projected through similar color filters and superimposed on the projection screen, an additive method of color reproduction. A color print on paper could be produced by superimposing carbon prints of the three images made in their complementary colors, a subtractive method of color reproduction pioneered by Louis Ducos du Hauron in the late 1860s. Russian photographer Sergei Mikhailovich Prokudin-Gorskii made extensive use of this color separation technique, employing a special camera which successively exposed the three color-filtered images on different parts of an oblong plate. Because his exposures were not simultaneous, unsteady subjects exhibited color "fringes" or, if rapidly moving through the scene, appeared as brightly colored ghosts in the resulting projected or printed images.

The development of color photography was held back by the limited sensitivity of early photographic materials, which were mostly sensitive to blue, only slightly sensitive to green and virtually insensitive to red. The discovery of dye sensitization by photochemist Hermann Vogel in 1873 suddenly made it possible to add sensitivity to green, yellow and even red. Improved color sensitizers and ongoing improvements in the overall sensitivity of emulsions steadily reduced the once-prohibitive long exposure times required for color, bringing it ever closer to commercial viability.

Autochrome, the first commercially successful color process, was introduced by the Lumière brothers in 1907. Autochrome plates incorporated a mosaic color filter layer made of dyed grains of potato starch, which allowed the three color components to be recorded as adjacent microscopic image fragments. After an Autochrome plate was reversal processed to produce a positive transparency, the starch grains served to illuminate each fragment with the correct color and the tiny colored points blended together in the eye, synthesizing the color of the subject by the additive method. Autochrome plates were one of several varieties of additive color screen plates and films marketed between the 1890s and the 1950s.

Kodachrome, the first modern "integral tripack" (or "monopack") color film, was introduced by Kodak in 1935. It captured the three color components in a multilayer emulsion. One layer was sensitized to record the red-dominated part of the spectrum, another layer recorded only the green part and a third recorded only the blue. Without special film processing, the result would simply be three superimposed black-and-white images, but complementary cyan, magenta, and yellow dye images were created in those layers by adding color couplers during a complex processing procedure. Agfa's similarly structured Agfacolor Neu was introduced in 1936. Unlike Kodachrome, the color couplers in Agfacolor Neu were incorporated into the emulsion layers during manufacture, which greatly simplified the processing. Currently available color films still employ a multilayer emulsion and the same principles, most closely resembling Agfa's product.

Instant color film, used in a special camera which yielded a unique finished color print only a minute or two after the exposure, was introduced by Polaroid in 1963.

Color photography may form images as positive transparencies, which can be used in a slide projector, or as color negatives intended for use in creating positive color enlargements on specially coated paper. The latter is now the most common form of film (non-digital) color photography owing to the introduction of automated photoprinting equipment.

Stereoscopic

Photographs, both monochrome and color, can be captured and displayed through two side-by-side images that emulate human stereoscopic vision. While known colloquially as "3-D" photography, the more accurate term is stereoscopy. Such cameras have long been realized by using film, and more recently in digital electronic methods (including cellphone cameras).

Full-spectrum, ultraviolet and infrared

Ultraviolet and infrared films have been available for many decades and employed in a variety of photographic avenues since the 1960s. New technological trends in digital photography have opened a new direction in full spectrum photography, where careful filtering choices across the ultraviolet, visible and infrared lead to new artistic visions.

Modified digital cameras can detect some ultraviolet, all of the visible and much of the near infrared spectrum, as most digital imaging sensors are sensitive from about 350 nm to 1000 nm. An off-the-shelf digital camera contains an infrared hot mirror filter that blocks most of the infrared and a bit of the ultraviolet that would otherwise be detected by the sensor, narrowing the accepted range from about 400 nm to 700 nm.[23] Replacing a hot mirror or infrared blocking filter with an infrared pass or a wide spectrally transmitting filter allows the camera to detect the wider spectrum light at greater sensitivity. Without the hot-mirror, the red, green and blue (or cyan, yellow and magenta) colored micro-filters placed over the sensor elements pass varying amounts of ultraviolet (blue window) and infrared (primarily red and somewhat lesser the green and blue micro-filters).

Uses of full spectrum photography are for fine art photography, geology, forensics and law enforcement.

Digital photography

A handheld digital camera, Canon Ixus class.

Traditional photography made it difficult for photographers who worked at remote locations without easy access to processing facilities, and competition from television pressured photographers to deliver images to newspapers with greater speed. Photojournalists at remote locations often carried miniature photo labs and a means of transmitting images through telephone lines. In 1981, Sony unveiled the first consumer camera to use a charge-coupled device for imaging, eliminating the need for film: the Sony Mavica. While the Mavica saved images to disk, the images were displayed on television, and the camera was not fully digital. In 1991, Kodak unveiled the DCS 100, the first commercially available digital single lens reflex camera. Although its high cost precluded uses other than photojournalism and professional photography, commercial digital photography was born.

The Nikon D1, the first DSLR to truly compete with, and begin to replace, film cameras in the professional photojournalism and sports photography fields.

Digital imaging uses an electronic image sensor to record the image as a set of electronic data rather than as chemical changes on film. The primary difference between digital and chemical photography is that chemical photography resists photo manipulation because it involves film and photographic paper, while digital imaging is a highly manipulative medium. This difference allows for a degree of image post-processing that is comparatively difficult in film-based photography and permits different communicative potentials and applications.

Nikon DSLR and scanner, which converts film images to digital

Digital imaging has raised ethical concerns because of the ease of manipulating digital photographs in post-processing. Many photojournalists have declared they will not crop their pictures, or are forbidden from combining elements of multiple photos to make "photomontages", passing them as "real" photographs. Today's technology has made image editing relatively simple for even the novice photographer. However, recent changes of in-camera processing allows digital fingerprinting of photos to detect tampering for purposes of forensic photography.

Digital point-and-shoot cameras have become widespread consumer products, outselling film cameras, and including new features such as video and audio recording. Kodak announced in January 2004 that it would no longer sell reloadable 35 mm cameras in western Europe, Canada and the United States after the end of that year. Kodak was at that time a minor player in the reloadable film cameras market. In January 2006, Nikon followed suit and announced that they will stop the production of all but two models of their film cameras: the low-end Nikon FM10, and the high-end Nikon F6. On May 25, 2006, Canon announced they will stop developing new film SLR cameras.[24] Though most new camera designs are now digital, a new 6x6cm/6x7cm medium format film camera was introduced in 2008 in a cooperation between Fuji and Voigtländer.[25] [26]

According to a survey made by Kodak in 2007 when the majority of photography was already digital, 75 percent of professional photographers say they will continue to use film, even though some embrace digital.[27]

According to the U.S. survey results, more than two-thirds (68 percent) of professional photographers prefer the results of film to those of digital for certain applications including:

- film's superiority in capturing more information on medium and large format films (48 percent);
- creating a traditional photographic look (48 percent);
- capturing shadow and highlighting details (45 percent);
- the wide exposure latitude of film (42 percent); and
- archival storage (38 percent)

Sony Ericsson K800i. Such Camera phones, combined with photo sharing sites, have led to a new kind of social photography.

Manual shutter control and exposure settings can achieve unusual results.

"Light field photography"

Digital methods of image capture and display processing have enabled the new technology of "light field photography" (also known as synthetic aperture photography). This process allows focusing at various depths of field to be selected *after* the photograph has been captured.[28] As explained by Michael Faraday in 1846, the "light field" is understood as 5-dimensional, with each point in 3-d space having attributes of two more angles that define the direction of each ray passing through that point. These additional vector attributes can be captured optically through the use of microlenses at each pixel-point within the 2-dimensional image sensor. Every pixel of the final image is actually a selection from each sub-array located under each microlens, as identified by a post-image capture focus algorithm.

Modes of production

Amateur

An amateur photographer is one who practices photography as a hobby and not for profit. The quality of some amateur work is comparable to that of many professionals and may be highly specialized or eclectic in its choice of subjects. Amateur photography is often pre-eminent in photographic subjects which have little prospect of commercial use or reward.

Commercial

Commercial photography is probably best defined as any photography for which the photographer is paid for images rather than works of art. In this light, money could be paid for the subject of the photograph or the photograph itself. Wholesale, retail, and professional uses of photography would fall under this definition. The commercial photographic world could include:

- Advertising photography: photographs made to illustrate and usually sell a service or product. These images, such as packshots, are generally done with an advertising agency, design firm or with an in-house corporate design team.
- Fashion and glamour photography usually incorporates models and is a form of advertising photography. Fashion photography, like the work featured in Harper's Bazaar, emphasizes clothes and other products; glamour emphasizes the model and body form. Glamour photography is popular in advertising and men's magazines. Models in glamour photography sometimes work nude.
- Crime scene photography consists of photographing scenes of crime such as robberies and murders. A black and white camera or an infrared camera may be used to capture specific details.
- Still life photography usually depicts inanimate subject matter, typically commonplace objects which may be either natural or man-made. Still life is a broader category for food and some natural photography and can be used for advertising purposes.
- Food photography can be used for editorial, packaging or advertising use. Food photography is similar to still life photography, but requires some special skills.
- Editorial photography illustrates a story or idea within the context of a magazine. These are usually assigned by the magazine and encompass fashion and glamour photography features.
  - Photojournalism can be considered a subset of editorial photography. Photographs made in this context are accepted as a documentation of a news story.
- Portrait and wedding photography: photographs made and sold directly to the end user of the images.
- Landscape photography depicts locations.
- Wildlife photography demonstrates the life of animals.
- Paparazzi is a form of photojournalism in which the photographer captures candid images of athletes, celebrities, politicians, and other prominent people.

Landscape 360-degree panoramic picture of the Chajnantor plateau. In the center is Cerro Chajnantor itself. To the right, on the plateau, is the Atacama Pathfinder Experiment (APEX) telescope with Cerro Chascon behind it.[29]

The market for photographic services demonstrates the aphorism "A picture is worth a thousand words", which has an interesting basis in the history of photography. Magazines and newspapers, companies putting up Web sites, advertising agencies and other groups pay for photography.

Many people take photographs for self-fulfillment or for commercial purposes. Organizations with a budget and a need for photography have several options: they can employ a photographer directly, organize a public competition, or obtain rights to stock photographs. Photo stock can be procured through traditional stock giants, such as Getty Images or Corbis; smaller microstock agencies, such as Fotolia; or web marketplaces, such as Cutcaster.

Art

Classic Alfred Stieglitz photograph, *The Steerage* shows unique aesthetic of black-and-white photos.

During the 20th century, both fine art photography and documentary photography became accepted by the English-speaking art world and the gallery system. In the United States, a handful of photographers, including Alfred Stieglitz, Edward Steichen, John Szarkowski, F. Holland Day, and Edward Weston, spent their lives advocating for photography as a fine art. At first, fine art photographers tried to imitate painting styles. This movement is called Pictorialism, often using soft focus for a dreamy, 'romantic' look. In reaction to that, Weston, Ansel Adams, and others formed the Group f/64 to advocate 'straight photography', the photograph as a (sharply focused) thing in itself and not an imitation of something else.

The aesthetics of photography is a matter that continues to be discussed regularly, especially in artistic circles. Many artists argued that photography was the mechanical reproduction of an image. If photography is authentically art, then photography in the context of art would need redefinition, such as determining what component of a photograph makes it beautiful to the viewer. The controversy began with the earliest images "written with light"; Nicéphore Niépce, Louis Daguerre, and others among the very earliest photographers were met with acclaim, but some questioned if their work met the definitions and purposes of art.

Clive Bell in his classic essay *Art* states that only "significant form" can distinguish art from what is not art.

> There must be some one quality without which a work of art cannot exist; possessing which, in the least degree, no work is altogether worthless. What is this quality? What quality is shared by all objects that provoke our aesthetic emotions? What quality is common to Sta. Sophia and the windows at Chartres, Mexican sculpture, a Persian bowl, Chinese carpets, Giotto's frescoes at Padua, and the masterpieces of Poussin, Piero della Francesca, and Cezanne? Only one answer seems possible — significant form. In each, lines and colors combined in a particular way, certain forms and relations of forms, stir our aesthetic emotions.
> __[30]

On February 14, 2006, Sotheby's London sold the 2001 photograph "99 Cent II Diptychon" for an unprecedented $3,346,456 to an anonymous bidder, making it the most expensive of all time.

- Conceptual photography

  Photography that turns a concept or idea into a photograph. Even though what is depicted in the photographs are real objects, the subject is strictly abstract.

Science and forensics

The camera has a long and distinguished history as a means of recording phenomena from the first use by Daguerre and Fox-Talbot, such as astronomical events (eclipses for example), small creatures and plants when the camera was attached to the eyepiece of microscopes (in photomicroscopy) and for macro photography of larger specimens. The camera also proved useful in recording crime scenes and the scenes of accidents, such as the Wootton bridge collapse in 1861. The methods used in analysing photographs for use in legal cases are collectively known as forensic photography.

Wootton bridge collapse in 1861

By 1853, Charles Brooke had invented a technology for the automatic registration of instruments by photography. These instruments included barometers, thermometers, psychrometers, and magnetometers, which recorded their readings by means of an automated photographic process.[31]

5×7 inch unretouched photograph of the Wright brothers' first flight, 1903

Photography has become ubiquitous in recording events and data in science and engineering, and at crime scenes or accident scenes. The method has been much extended by using other wavelengths, such as infrared photography and ultraviolet photography, as well as spectroscopy. Those methods were first used in the Victorian era and developed much further since that time.[32]

Other image forming techniques

Besides the camera, other methods of forming images with light are available. For instance, a photocopy or xerography machine forms permanent images but uses the transfer of static electrical charges rather than photographic film, hence the term electrophotography. Photograms are images produced by the shadows of objects cast on the photographic paper, without the use of a camera. Objects can also be placed directly on the glass of an image scanner to produce digital pictures.

Social and cultural implications

There are many ongoing questions about different aspects of photography. In her writing "On Photography" (1977), Susan Sontag discusses concerns about the objectivity of photography. This is a highly debated subject within the photographic community.[33] Sontag argues, "To photograph is to appropriate the thing photographed. It means putting one's self into a certain relation to the world that feels like knowledge, and therefore like power."[34] Photographers decide what to take a photo of, what elements to exclude and what angle to frame the photo, and these factors may reflect a particular socio-historical context. Along these lines it can be argued that photography is a subjective form of representation.

Modern photography has raised a number of concerns on its impact on society. In Alfred Hitchcock's *Rear Window* (1954), the camera is presented as promoting voyeurism. 'Although the camera is an observation station, the act of photographing is more than passive observing'.[34] Michal Powell's *Peeping Tom* (1960) portrays the camera as both sexual and sadistically violent technology that literally kills in this picture and at the same time captures images of the pain and anguish evident on the faces of the female victims.

> "The camera doesn't rape or even possess, though it may presume, intrude, trespass, distort, exploit, and, at the farthest reach of metaphor, assassinate - all activities that, unlike the sexual push and shove, can

be conducted from a distance, and with some detachment."[34]

Photography is one of the new media forms that changes perception and changes the structure of society.[35] Further unease has been caused around cameras in regards to desensitization. Fears that disturbing or explicit images are widely accessible to children and society at large have been raised. Particularly, photos of war and pornography are causing a stir. Sontag is concerned that "to photograph is to turn people into objects that can be symbolically possessed." Desensitization discussion goes hand in hand with debates about censored images. Sontag writes of her concern that the ability to censor pictures means the photographer has the ability to construct reality.[34]

One of the practices through which photography constitutes society is tourism. Tourism and photography combine to create a "tourist gaze"[36] in which local inhabitants are positioned and defined by the camera lens. However, it has also been argued that there exists a "reverse gaze"[37] through which indigenous photographees can position the tourist photographer as a shallow consumer of images.

Additionally, photography has been the topic of many songs in popular culture.

Law

Photography is both restricted and protected by the law in many jurisdictions. Protection of photographs is typically achieved through the granting of copyright or moral rights to the photographer. In the UK a recent law (Counter-Terrorism Act 2008) increases the power of the police to prevent people, even press photographers, from taking pictures in public places.[38]

See also

Forms

- Aviation photography
- Architectural photography
- Candid photography
- Cloudscape photography
- Conservation photography
- Cosplay photography
- Digiscoping
- Documentary photography
- Erotic photography
- Fashion photography
- Fine art photography
- Fire photography
- Food photography
- Forensic photography
- Glamour photography
- Head shot
- Landscape art
- Landscape photography
- Medical photography
- Miksang (contemplative photography)
- Nature photography
- Social photography
- Nude photography
- Old-time photography

- Photojournalism
- Portrait photography
- Sports photography
- Still life photography
- Stock photography
- Street photography
- Travel photography
- Underwater photography
- Vernacular photography
- VR photography
- War photography
- Wedding photography
- Wildlife photography

Photographers and photographs

- List of most expensive photographs
- List of photographers
- Movie stills photographer

Equipment (cameras, etc.)

- Camera
- Camera Phone
- Color chart
- Digital camera
- Digital single-lens reflex camera
- Dry box
- Film base
- Film format
- Film holder
- Film scanner
- Film stock
- Filter
- Flash
- Gray card
- Lenses for SLR and DSLR cameras
- List of photographic equipment makers
- Monopod
- Movie projector
- Perspective control lens
- Photographic film
- Photographic lens
- Reflector
- Rangefinder camera
- SD Card(for digital photography)
- Single-lens reflex camera
- Slide projector
- Soft box
- Still camera

- Toy camera
- Tripod
- Twin-lens reflex camera
- Video camera
- View camera
- Zone plate

History

- Albumen print
- Calotype
- Daguerreotype
- Timeline of photography technology

Techniques

- Aerial Photography
- Afocal photography
- Astrophotography
- Bokeh
- Contre-jour
- Cross processing
- Cyanotype
- Fill flash
- Film developing
- Full spectrum photography
- Harris Shutter
- High dynamic range imaging
- High speed photography
- Image fusion
- Infrared photography
- Kinetic photography
- Kite aerial photography
- Lead room
- Light painting
- Lith-Print
- Macro photography
- Micrography, or Photomicrography
- Monochrome Photography
- Motion blur
- Night photography
- Panning
- Panoramic photography
- Photogram
- Photograph conservation
- Photographic mosaic
- Photographic print toning
- Push printing
- Push processing
- Rephotography

- Rollout photography
- Sabatier Effect
- Schlieren photography
- Stereoscopy
- Sun printing
- Tilted plane focus
- Time-lapse
- Ultraviolet photography
- Wide dynamic range
- Zoom burst

General concepts

- Adobe Photoshop
- Camera obscura
- Composition in visual arts
- Diana camera
- Early photographers of York
- Gelatin-silver process
- Gum printing
- Hand-coloring
- Holography
- Kirlian photography
- Lomography
- Mourning portraits
- Negative
- North American Nature Photography Association
- Photograph
- Print permanence
- Vignetting

Technical principles

- Angle of view
- Aperture
- Color temperature
- Depth of field
- Depth of focus
- Digital versus film photography
- Double exposure
- Exposure
- F-number
- Film format
- Film speed
- Perspective distortion
- Photographic printing
- Photographic processes
- Pinhole camera
- Reciprocity (photography)
- Red-eye effect

- Rule of thirds
- Science of photography
- Shutter speed
- Zone System

References

[1] Spencer, D A (1973). *The Focal Dictionary of Photographic Technologies*. Focal Press. p. 454. ISBN 240 50747 9.
[2] Eder, J.M (1945) [1932]. *History of Photography, 4th. edition [Geschichte der Photographie]*. New York: Dover Publications, Inc.. pp. 258–259. ISBN 0486235866.
[3] φάος (http://www.perseus.tufts.edu/hopper/text?doc=Perseus:text:1999.04.0057:entry=fa/os), Henry George Liddell, Robert Scott, *A Greek-English Lexicon*, on Perseus
[4] γραφή (http://www.perseus.tufts.edu/hopper/text?doc=Perseus:text:1999.04.0057:entry=grafh/), Henry George Liddell, Robert Scott, *A Greek-English Lexicon*, on Perseus
[5] Online Etymology Dictionary (http://www.etymonline.com/index.php?search=photography&searchmode=none)
[6] Dpreview.com (http://www.dpreview.com/learn/?/glossary/)
[7] Joseph and Barbara Anderson, "The Myth of Persistence of Vision Revisited", *Journal of Film and Video*, Vol. 45, No. 1 (Spring 1993): 3–12. uca.edu (http://www.uca.edu/org/ccsmi/ccsmi/classicwork/Myth Revisited.htm)
[8] "Definition of focus" (http://dictionary.reference.com/browse/focus). IAC. . Retrieved 31 January 2012.
[9] "The First Photograph — Heliography" (http://www.hrc.utexas.edu/exhibitions/permanent/wfp/heliography.html). . Retrieved 2009-09-29. "from Helmut Gernsheim's article, "The 150th Anniversary of Photography," in History of Photography, Vol. I, No. 1, January 1977: ...In 1822, Niépce coated a glass plate... The sunlight passing through... This first permanent example... was destroyed... some years later."
[10] Jan Campbell (2005). " *Film and cinema spectatorship: melodrama and mimesis* (http://books.google.com/books?id=lOEqvkmSxhsC&pg=PA114&dq&hl=en#v=onepage&q=&f=false)". Polity. p. 114. ISBN 0-7456-2930-X
[11] Robert E. Krebs (2004). *Groundbreaking Scientific Experiments, Inventions, and Discoveries of the Middle Ages and the Renaissance* (http://books.google.com/?id=MTXdplfiz-cC&pg=PA20&dq=Mo-Ti+pinhole+camera+obscura). Greenwood Publishing Group. ISBN 0-3133-2433-6. .
[12] Alistair Cameron Crombie, *Science, optics, and music in medieval and early modern thought*, p. 205
[13] Wade, Nicholas J.; Finger, Stanley (2001). "The eye as an optical instrument: from camera obscura to Helmholtz's perspective". *Perception* **30** (10): 1157–77. doi:10.1068/p3210. PMID 11721819.
[14] Davidson, Michael W; National High Magnetic Field Laboratory at The Florida State University (2003-08-01). "Molecular Expressions: Science, Optics and You - Timeline - Albertus Magnus" (http://micro.magnet.fsu.edu/optics/timeline/people/magnus.html). The Florida State University. . Retrieved 2009-11-28.
[15] Georges Potonniée (1973). *The history of the discovery of photography*. Arno Press. p. 50. ISBN 0-4050-4929-3
[16] Helmut Gernsheim (1986). *A concise history of photography* (http://books.google.com/books?id=GDSRJQ3BZ5EC&pg=PA3&dq&hl=en#v=onepage&q=&f=false). Courier Dover Publications. pp. 3–4. ISBN 0-4862-5128-4
[17] Helmut Gernsheim, Alison Gernsheim (1955). "*The history of photography from the earliest use of the camera obscura in the eleventh century up to 1914*". Oxford University Press. p. 20.
[18] Witt, Brown, Dunbar, Tirro, Witt. The Humanities, Cultural Roots and Continuities, Seventh Edition. Houghton Mifflin Company. Boston. New York. 2005
[19] Seizing the Light: A History of Photography (http://books.google.com/books?id=vftTAAAAMAAJ&q=Joseph+Nicephore+Niepce+View+From+the+Window+of+Gras&dq=Joseph+Nicephore+Niepce+View+From+the+Window+of+Gras&client=safari&cd=3) By Robert Hirsch
[20] BBC - History - Historic Figures: William Henry Fox Talbot (1800 - 1877) (http://www.bbc.co.uk/history/historic_figures/fox_talbot_william_henry.shtml) BBC
[21] Anthony Feldman, Peter Ford (1989) *Scientists & inventors* p. 128. Bloomsbury Books, 1989
[22] *William H. Fox Talbot, inventor of the negative-positive process* p. 95. Macmillan, 1973
[23] Spectral curves of RGB and Hot Mirror filters. (http://surrealcolor.110mb.com/IR_explained_web/IR_explained.htm#CamColor)
[24] "Canon to Stop Making Single-Lens Camera" (http://www.indexstockimagery.com/archives/2006/05/canon_to_stop_m.html) Associated Press, 25 May 2006. Retrieved 2 September 2006.
[25] Voigtlaender.de (http://www.voigtlaender.de/cms/voigtlaender/voigtlaender_cms.nsf/id/pa_fdih7jzkae.html)
[26] The new Voigtlaender Vitolux S70 and Bessa III 667 (http://www.dcviews.com/press/Voigtlaender-Bessa.htm)
[27] www.photographypress.co.uk (http://www.photographypress.co.uk/news/news.phtml/6443/7467/Kodak-Survey-Photographers-Use-Film.phtml)
[28] Digital Light Field Photography, Ren Ng 2006 PhD dissertation (https://www.lytro.com/renng-thesis.pdf)
[29] "All Around Chajnantor — A 360-degree panorama" (http://www.eso.org/public/images/potw1215a/). *ESO Picture of the Week*. . Retrieved 13 April 2012.

[30] Clive Bell. " Art (http://www.csulb.edu/~jvancamp/361r13.html)", 1914. Retrieved 2 September 2006.

[31] "Photographic self-registering magnetic and meteorological apparatus: Invented by Mr. Brooke of Keppel-Street, London" (http://books.google.com/books?id=DhfnAAAAMAAJ&pg=PA309). *The Illustrated Magazine of Art* (New York: Alexander Montgomery) **1**: 308–311. 1853. .

[32] Herbert L. Blitzer, Karen Stein-Ferguson, Jeffrey Huang (2008). *Understanding forensic digital imaging* (http://books.google.com/books?id=a0nmdTmHMrIC&pg=PA8). Academic Press. pp. 8–9. ISBN 9780123704511. .

[33] Bissell, K.L., Photography and Objectivity (2000) findarticles.com (http://findarticles.com/p/articles/mi_qa3677/is_200007/ai_n8895320) . Retrieved 24 October 2008.

[34] Sontag, S., On Photography, Penguin, London (1977), pp 3–24.

[35] Levinson, P., The Soft Edge: a Natural History and Future of the Information Revolution, Routledge, London and New York (1997), pp 37–48.

[36] John Urry (2002). *The tourist gaze* (http://books.google.com/?id=bhhtg1sz0YAC&printsec=frontcover#v=onepage&q=) (2nd ed.). SAGE. ISBN 9780761973478. .

[37] Alex Gillespie. "Tourist Photography and the Reverse Gaze" (http://stir.academia.edu/documents/0011/0117/Gillespie_tourist_photography_and_the_reverse_gaze.pdf). .

[38] British Journal of Photography article (http://www.bjp-online.com/public/showPage.html?page=836675)

Further reading

Introduction

- *Photography. A Critical Introduction* [Paperback], ed. by Liz Wells, 3rd edition, London [etc.]: Routledge, 2004, ISBN 041530704X

History

- *A New History of Photography*, ed. by Michel Frizot, Köln : Könemann, 1998
- Franz-Xaver Schlegel, *Das Leben der toten Dinge - Studien zur modernen Sachfotografie in den USA 1914-1935*, 2 Bände, Stuttgart/Germany: Art in Life 1999, ISBN 3-00-004407-8.

Reference works

- Tom Ang (2002). *Dictionary of Photography and Digital Imaging: The Essential Reference for the Modern Photographer* (http://books.google.com/?id=fu3akyrFZEMC&pg=PP1&dq=intitle:Dictionary+intitle:of+intitle:Photography+intitle:and+intitle:Digital+intitle:Imaging+inauthor:ang). Watson-Guptill. ISBN 0817437894.
- Hans-Michael Koetzle: *Das Lexikon der Fotografen: 1900 bis heute*, Munich: Knaur 2002, 512 p., ISBN 3-426-66479-8
- John Hannavy (ed.): *Encyclopedia of Nineteenth-Century Photography*, 1736 p., New York: Routledge 2005 ISBN 978-0415972352
- Lynne Warren (Hrsg.): *Encyclopedia of Twentieth-Century Photography*, 1719 p., New York, NY [et.] : Routledge, 2006
- *The Oxford Companion to the Photograph*, ed. by Robin Lenman, Oxford University Press 2005

Other books

- *Photography and The Art of Seeing* by Freeman Patterson, Key Porter Books 1989, ISBN 1-55013-099-4.
- *The Art of Photography:* An Approach to Personal Expression by Bruce Barnbaum, Rocky Nook 2010, ISBN 1933952687.
- *Image Clarity: High Resolution Photography* by John B. Williams, Focal Press 1990, ISBN 0-240-80033-8.

External links

- Photography (http://www.dmoz.org/Arts/Photography/) at the Open Directory Project
- World History of Photography (http://all-art.org/history658_photography1.html) From The History of Art.
- Daguerreotype to Digital: A Brief History of the Photographic Process (http://www.floridamemory.com/photographiccollection/photo_exhibits/photographic-processes/) From the State Library & Archives of Florida.
- Shades of Light (Australian Photography 1839 - 1988) (http://www.photo-web.com.au/shadesoflight) the online version of the original Shades of Light published 1998, Gael Newton,

k h :

h n :

Visual_effects

Visual effects (commonly shortened to **Visual F/X** or **VFX**) are the various processes by which imagery is created and/or manipulated outside the context of a live action shoot. Visual effects involve the integration of live-action footage and generated imagery to create environments which look realistic, but would be dangerous, costly, or simply impossible to capture on film. Visual effects using computer generated imagery (CGI) have become increasingly common in big-budget films, and have also recently become accessible to the Independent filmmaker with the introduction of affordable animation and compositing software.

Timing

Visual effects are often integral to a movie's story and appeal. Although most visual effects work is completed during post-production, it usually must be carefully planned and choreographed in pre-production and production. Visual effects are designed and edited in Post-Production, with the use of graphic design, modeling, animation and similar software, while special effects are made on set, such as explosions, car chases and so on. A visual effects supervisor is usually involved with the production from an early stage to work closely with production and the film's director to achieve the desired effects.

Categories

Visual effects may be divided into at least four categories:

- Models: miniature sets and models, animatronics, stop motion animation.
- Matte paintings and stills: digital or traditional paintings or photographs which serve as background plates for keyed or rotoscoped elements.
- Live-action effects: keying actors or models through bluescreening and greenscreening.
- Digital animation: modeling, computer graphics lighting, texturing, rigging, animating, and rendering computer-generated 3D characters, particle effects, digital sets, backgrounds.
- **Digital effects** (commonly shortened to **digital FX** or **FX**) are the various processes by which imagery is created and/or manipulated with or from photographic assets. Digital effects often involve the integration of still photography and computer generated imagery (CGI) in order to create environments which look realistic, but

would be dangerous, costly, or simply impossible to capture in camera. FX is usually associated with the still photography world in contrast to visual effects which is associated with motion film production.
- Amateur VFX Animation which are VFX made by budding VFX artists. e.g. Fancock 4 and LTS

See also

- Animation
- Bluescreen
- Compositing
- Computer-generated imagery
- Front projection effect
- Matte painting
- Physical effects, another category of special effects
- Optics#Visual effects
- Rear projection effect
- VFX Creative Director

Further reading

- The VES Handbook of Visual Effects: Industry Standard VFX Practices and Procedures, Jeffrey A. Okun & Susan Zwerman, Publisher: Focal Press 2010
- T. Porter and T. Duff, "Compositing Digital Images", Proceedings of SIGGRAPH '84, 18 (1984).
- The Art and Science of Digital Compositing (ISBN 0-12-133960-2)
- McClean, Shilo T. (2007). *Digital Storytelling: The Narrative Power of Visual Effects in Film*. The MIT Press. ISBN 0-262-13465-9.
- Mark Cotta Vaz; Craig Barron: *The Invisible Art: The Legends of Movie Matte Painting*. San Francisco, Cal.: Chronicle Books, 2002; ISBN 0-8118-3136-1
- Peter Ellenshaw; *Ellenshaw Under Glass - Going to the Matte for Disney* [1]
- Richard Rickitt: *Special Effects: The History and Technique*. Billboard Books; 2nd edition, 2007; ISBN 0823084086.
- Patel, Mayur (2009). *The Digital Visual Effects Studio: The Artists and Their Work Revealed*. ISBN 1-448-66547-7.

External links

- Take Five Minutes to Watch 100 Years of Visual Effects [2] by Rosa Golijan - Gizmodo.com - August 27, 2009

References

[1] http://www.ellenshaw.com/bk_Under_Glass.htm
[2] http://gizmodo.com/5347365/take-five-minutes-to-watch-100-years-of-visual-effects?autoplay=true

Slow_motion

Slow motion (commonly abbreviated as **slowmo**) is an effect in film-making whereby time appears to be slowed down. It was invented by the Austrian priest August Musger.

Typically this style is achieved when each film frame is captured at a rate much faster than it will be played back. When replayed at normal speed, time appears to be moving more slowly. The technical term for slow motion is **overcranking** which refers to the concept of cranking a handcranked camera at a faster rate than normal (i.e. faster than 24 frames per second). Slow motion can also be achieved by playing normally recorded footage at a slower speed. This technique is more often applied to video subjected to instant replay, than to film. A third technique that is becoming common using current computer software post-processing (with programs like Twixtor) is to fabricate digitally interpolated frames to smoothly transition between the frames that were actually shot. Motion can be slowed further by combining techniques, interpolating between overcranked frames. The traditional method for achieving super-slow motion is through high-speed photography, a more sophisticated technique that uses specialized equipment to record fast phenomena, usually for scientific applications.

Slow motion is ubiquitous in modern filmmaking. It is used by a diverse range of directors to achieve diverse effects. Some classic subjects of slow motion include:

- Athletic activities of all kinds, to demonstrate skill and style.
- To recapture a key moment in an athletic game, typically shown as a replay.
- Natural phenomena, such as a drop of water hitting a glass.

Slow motion can also be used for artistic effect, to create a romantic or suspenseful aura or to stress a moment in time. Vsevolod Pudovkin, for instance, used slow motion in a suicide scene in *The Deserter*, in which a man jumping into a river seems sucked down by the slowly splashing waves. Another example is *Face/Off*, in which John Woo used the same technique in the movements of a flock of flying pigeons. *The Matrix* made a distinct success in applying the effect into action scenes through the use of multiple cameras, as well as mixing slow-motion with live action in other scenes. Japanese director Akira Kurosawa was a pioneer using this technique in his 1954 movie *Seven Samurai*. American director Sam Peckinpah was another classic lover of the use of slow motion. The technique is especially associated with explosion effect shots and underwater footage.

The opposite of slow motion is fast motion. Cinematographers refer to fast motion as **undercranking** since it was originally achieved by cranking a handcranked camera slower than normal. It is often used for comic effect, time lapse or occasional stylistic effect.

The concept of slow motion may have existed before the invention of the motion picture: the Japanese theatrical form Noh employs very slow movements.

How slow motion works

There are two ways in which slow motion can be achieved in modern cinematography. Both involve a camera and a projector. A projector refers to a classical film projector in a movie theater, but the same basic rules apply to a television screen and any other device that displays consecutive images at a constant frame rate.

Overcranking

For the purposes of making the above illustration readable a projection speed of 10 frames per second (frame/s) has been selected, in fact film is usually projected at 24 frame/s making the equivalent slow overcranking is rare, but available on professional equipment.[1]

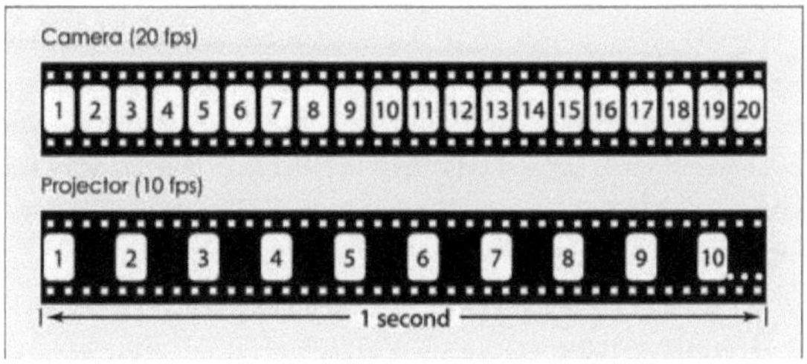

Time stretching

The second type of slow motion is achieved during post production. This is known as *time-stretching* or *digital slow motion*. This type of slow motion is achieved by inserting new frames in between frames that have actually been photographed. The effect is similar to overcranking as the actual motion occurs over a longer time.

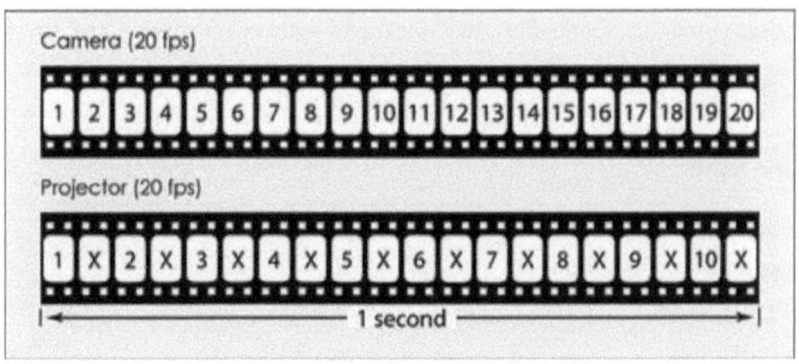

Frames marked with an X must be fabricated.

Since the necessary frames were never photographed, new frames must be fabricated. Sometimes the new frames are simply repeats of the preceding frames but more often they are created by interpolating between frames. (Often this interpolation is effectively a short dissolve between still frames). Many complicated algorithms exist that can track motion between frames and generate intermediate frames that appear natural and smooth. However it is understood that these methods can never achieve the clarity or smoothness of its overcranking counterpart.

Traditionally, frames were duplicated on an optical printer. True frame interpolation can only be done digitally.

Simple replication of the same frame twice is also sometimes called *half-speed*. This relatively primitive technique (as opposed to digital interpolation) is often visually detectable by the casual viewer. It was used in certain scenes in *Tarzan, the Ape Man*, and critics pointed it out. Sometimes lighting limitations or editorial decisions can require it. A wide-angle shot of Roy Hobbs swinging the bat, in the climactic moments of *The Natural*, was printed at half-speed in order to simulate slow-motion, and the closeup that immediately followed it was true overcranked slow-motion.

A VCR may have the option of slow motion playback, sometimes at various speeds; this can be applied to any normally recorded scene. It is similar to half-speed, and is not true slow-motion, but merely longer display of each frame.

In action films

Slow motion is used widely in action films for dramatic effect, as well as the famous bullet-dodging effect, popularized by *The Matrix*.

Formally, this effect is referred to as **speed ramping** and is a process whereby the capture frame rate of the camera changes over time. For example, if in the course of 10 seconds of capture, the capture frame rate is adjusted from 60 frames per second to 24 frames per second, when played back at the standard film rate of 24 frames per second, a unique time-manipulation effect is achieved. For example, someone pushing a door open and walking out into the street would appear to start off in slow-motion, but in a few seconds later within the same shot the person would appear to walk in "realtime" (normal speed). The opposite speed-ramping is done in *The Matrix* when Neo re-enters the Matrix for the first time to see the Oracle. As he comes out of the warehouse "load-point", the camera zooms into Neo at normal speed but as it gets closer to Neo's face, time seems to slow down, perhaps visually accentuating Neo pausing and reflecting a moment, and perhaps alluding to future manipulation of time itself within the Matrix later on in the movie.

In Broadcasting

Slow-motion is widely used in sport broadcasting and its origins in this domain extend right back to the earliest days of television, one example being the European Heavyweight Title in 1939 where Max Schmeling knocked out Adolf Heuser in 71 seconds.[2]

In instant replays, slow motion reviews are now commonly used to show in detail some action (photo finish, Football (soccer) goal, ...). Generally, they are made with video servers and special controllers. The first TV slo-mo was the Ampex HS-100 disk record-player.

See also

- Motion picture terminology
- High speed camera
- Time-lapse
- Bullet time
- Video server

HS-100 at DC Video, [3],

- Multicam (LSM)
- Temporal posterization
- Maurice Clarett

References

[1] (http://www.sony.co.uk/biz/view/ShowProduct.action?product=F23&site=biz_en_GB&pageType=Features&category=HDCAMSRCamcorders)

[2] Kloft, Michael (Director) (1999). *Das Fernsehen unter dem Hakenkreuz* (http://www.imdb.com/title/tt0256749/) (Documentary). Germany: Spiegel TV. .

[3] http://www.dcvideo.com/

External links

- Interesting High-speed Video Clips (http://www.engr.colostate.edu/~dga/high_speed_video/)
- Lucid Movement, high-speed daily video blog (http://www.lucidmovement.com)
- JackCabbage: Overcrank on the EX-1 (http://jackcabbage.blogspot.com/2008/06/sony-ex1-overcrank-footage.html)
- JackCabbage: Overcrank and Undercrank on the HVX (http://jackcabbage.blogspot.com/2008/06/hvx-overcrank-and-undercrank.html)

Camera_angle

The **camera angle** marks the specific location at which a camera is placed to take a shot. A scene may be shot from several camera angles.[1] This will give different experience and sometimes emotion. the different camera angles will have different effects on the viewer and how they perceive the scene that is shot. There are a few different routes that a camera operator could take to achieve their desired effect.

Angles and their effects

Where the camera is placed in relation to the subject greatly effects the way the viewer perceives that subject. A Viewpoint is the apparent distance and angle from which the camera views and records the subject.[2] There are many different types of camera angles such as the high-angle shot, the low-angle shot, the bird's eye view and the worm's eye view.

They also include the eye level shot and the point of view shot. A high-angle shot (HA) is a shot in which the camera is physically higher than the subject and is looking down upon the subject. The high angle shot can make the subject look small or weak or vulnerable while a low-angle shot (LA) is taken from below the subject and has the power to make the subject look powerful or threatening. A neutral shot or eye Level (EL) shot has little to no psychological effect on the viewer. This shot is when the camera is level or looking straight on with the subject.

A point of view (POV) shot shows the audience the image through the subject's eye. Some POV shots use hand-held cameras to create the illusion that the audience is seeing through the subject's eyes. bird's eye shot or bird's-eye view[3] shots are taken directly above the scene to establish the landscape and the actors relationship to it. worm's eye shot or worm's eye view is a shot that is looking up from the ground, and is meant to give the viewer the feeling that they are looking up at the character from way below and it is meant to show the view that a child or a pet would have. When considering the camera angle one must remember that each shot is its own individual shot and the camera angle should be taken in context of the scene and film.

There are many different types of shots that can be used from these angles. There are extreme long shots which are extremely far away from the subject and might not even show a person at all. Extreme long shots are usually done in a high angle so the audience can look down upon a setting or scene. Extreme longs shots are used mainly to open the scene or narrative and show the audience the setting. The rest of the shots are most typically done in an eye level or point of view shot although it is possible to do any shot with any angle. There is the long shot which shows the subject even though the setting still dominates the picture frame.

Then. there is the medium long shot which makes the subject and the setting have equal importance and has the two about 50/50 in the frame. Then is the medium shot which emphasizes the character and is about a knees to waist up type shot. Then the medium close up is a shot that has the waist to the chest and up. The next closest shot is the close up which has the shoulders and up or maybe a little tighter on the head.

Finally, there is the extreme close up shot which has one body part usually. This can be an eye, a hand or anything else. These shots can be used with any of the aforementioned camera angles.[4]

A dutch angle, which can also be called a canted angle or even simply the tilted angle, is an angle in which the camera itself is tilted to the left or the right. The unnatural angle gives the viewer a feeling that world is out of balance or psychological unrest.

Naming a camera setup

Early example of use of camera angle

During production and post-production, it is necessary to give a unique alphanumeric identity to each camera angle, labeled as "scenes."[1] For example: "Scene 24C." Camera angle letters are often pronounced on the set using either the NATO phonetic alphabet or the older police-style radio alphabet. For example: "Scene 24C" would be pronounced as "Scene 24, Charlie." Some letters are skipped because they look like letters or numbers when written (for example an "S" can look like a "5").

See also

- Aerial perspective
- Aerial shot
- American shot
- Angle of view
- Bird's eye shot
- Bird's-eye view
- Boom shot
- B-roll
- Camera coverage
- Camera Dolly
- Camera operator
- Camera tracking
- Cinematic techniques
- Close-up
- Crane shot
- Dolly zoom
- Dutch angle
- Establishing shot
- Film frame
- Filmmaking
- Follow shot
- Forced perspective
- Freeze frame shot
- Full frame
- Full shot
- Hanging miniature
- Head shot
- High-angle shot
- Long shot
- Long take
- Low-angle shot
- Master shot
- Medium shot
- Money shot
- Multiple-camera setup
- One shot (music video)
- Over the shoulder shot
- Panning (camera)
- Point of view shot
- Rack focus
- Reaction shot
- Shaky camera
- Shot (filmmaking)
- Shot reverse shot
- Single-camera setup
- Tilt (camera)
- Top-down perspective
- Tracking shot
- Trunk shot
- Two shot
- Video production
- Walk and talk
- Whip pan
- Worm's-eye view

References

[1] Ascher, Steven, and Edward Pincus. *The Filmmaker's Handbook: A Comprehensive Guide for the Digital Age*. New York: Plume, 1999. p. 214

[2] Chandler, Daniel. "Grammar of Television and Film" (http://www.tlu.ee/~hagi/introtv2010/The grammar of TV - by D.Chandler.doc). .

[3] http://www.videomaker.com/article/9128/

[4] Chandler, Daniel. "The Grammar of TV Film" (http://www.aber.ac.uk/media/Documents/short/gramtv.html). . . Retrieved 14 November 2011.

Morphing

Morphing is a special effect in motion pictures and animations that changes (or morphs) one image into another through a seamless transition. Most often it is used to depict one person turning into another through technological means or as part of a fantasy or surreal sequence. Traditionally such a depiction would be achieved through cross-fading techniques on film. Since the early 1990s, this has been replaced by computer software to create more realistic transitions.

Three frames form a morph from George W. Bush to Arnold Schwarzenegger showing the mid-point between the two extremes

Early examples of morphing

Though the 1986 movie *The Golden Child* implemented very crude morphing effects from animal to human and back, the first movie to employ detailed morphing was *Willow*, in 1988. A similar process was used a year later in *Indiana Jones and the Last Crusade* to create Walter Donovan's gruesome demise. Both effects were created by Industrial Light & Magic using grid warping techniques developed by Tom Brigham and Doug Smythe (AMPAS).[1]

In 1985, Godley & Creme created a primitive "morph" effect using analogue cross-fades in the video for "Cry". The cover for Queen's 1989 album *The Miracle* featured the technique to morph the four band members' faces into one gestalt image. In 1991, morphing appeared notably in the Michael Jackson music video *Black or White* and in the movies *Terminator 2: Judgment Day* and *Star Trek VI: The Undiscovered Country*. The first application for personal computers to offer morphing was Gryphon Software Morph on the Macintosh. Other early morphing systems included ImageMaster, MorphPlus and CineMorph, all of which premiered for the Commodore Amiga in 1992. Other programs became widely available within a year, and for a time the effect became common to the point of cliché. For high-end use, Elastic Reality (based on MorphPlus) saw its first feature film use in *In The Line of Fire* (1993) and was used in Quantum Leap (work performed by the Post Group). At VisionArt Ted Fay used Elastic Reality to morph Odo for *Star Trek: Deep Space Nine*. Elastic Reality was later purchased by Avid, having already become the de facto system of choice, used in many hundreds of films. The technology behind Elastic Reality earned two Academy Awards in 1996 for Scientific and Technical Achievement going to Garth Dickie and Perry Kivolowitz. The effect is technically called a "spatially warped cross-dissolve". The first social network designed for user-generated morph examples to be posted online was Galleries by Morpheus (morphing software).

In Taiwan, Aderans, a hair loss solutions provider, did a TV commercial featuring a morphing sequence in which people with lush, thick hair morph into one another, reminiscent of the end sequence of the Black or White video.

Modern morphing techniques

Computer-animated morphing was used in the 1974 canadian animation Hunger.

In the early 1990s computer techniques that often produced more convincing results began to be widely used. These involved distorting one image at the same time that it faded into another through marking corresponding points and vectors on the "before" and "after" images used in the morph. For example, one would morph one face into another by marking key points on the first face, such as the contour of the nose or location of an eye, and mark where these same points existed on the second face. The computer would then distort the first face to have the shape of the second face at the same time that it faded the two faces.

Later, more sophisticated cross-fading techniques were employed that vignetted different parts of one image to the other gradually instead of transitioning the entire image at once. This style of morphing was perhaps most famously employed in the video that former 10cc members Kevin Godley and Lol Creme (performing as Godley & Creme) produced in 1985 for their song *Cry*. It comprised a series of black and white close-up shots of faces of many different people that gradually faded from one to the next. In a strict sense, this had little to do with modern-day computer generated morphing effects, since it was merely a dissolve using fully analog equipment.

Morphing in the future

Morphing algorithms continue to advance today and many programs can automatically morph images that correspond closely enough with relatively little instruction from the user. This has led to the use of morphing techniques to create convincing slow-motion effects where none existed in the original film or video footage by morphing between each individual frame using optical flow technology. Morphing has also appeared as a transition technique between one scene and another in television shows, even if the contents of the two images are entirely unrelated. The algorithm in this case attempts to find corresponding points between the images and distort one into the other as they crossfade.

Morphing is used far more heavily today than ever before. In years past, effects were obvious, which led to their overuse. Now, morphing effects are most often designed to be seamless and invisible to the eye.

Software

- Gryphon Software Morph
- Morpheus (morphing software)
- MorphThing

See also

- Morph target animation
- Inbetweening

References

[1] AMPAS - Index of Motion Picture Credits - Films (http://wwwdb.oscars.org/ampas/plsql/ampas_search.main?free_text_in=TOM BRIGHAM)

External links

- Tutorial on morphing using Adobe After Effects (http://www.creativecow.net/articles/zwar_chris/morph/index.html)
- Morph images on Mac OS X (http://www.norrkross.com/software/morphx/morphx.php)
- Xmorph: A Walkthrough of Morphing (http://xmorph.sourceforge.net/)

Virtual_cinematography

Virtual cinematography is an umbrella term used to describe cinematographic techniques performed in a computer graphics environment. This includes a wide variety of subjects like photographing real objects for the purpose of recreating them as three dimensional objects or algorithms for automated creation of camera viewpoints.

History

The term "virtual cinematography" came into the widespread usage following the release of the *The Matrix* films. The directors, Andy & Larry Wachowski, tasked visual effects supervisor John Gaeta with developing techniques to allow for virtual "filming" of realistic computer-generated imagery. Gaeta, along with Kim Libreri and his crew at ESC Entertainment succeeded, where many others failed, in creating photo-realistic CGI versions of performers, sets, and action.

Methods

Once the virtual content has been assembled into a scene within a 3D engine, it can then be creatively composed and re-photographed from other angles by a virtual camera as if the action was happening for the first time, hence the term "virtual cinematography". Modification, re-direction and enhancements to the scene are possible as well. The rendered result can appear highly realistic, or rather, "photo-realistic". Virtual cinematography is the creation process. Virtual effects are stylistic modifications being applied within this format. Virtual cinema is the result.

The art of "photographing" any computer-generated imagery content is becoming generally referred to as virtual cinematography. In both cases the term serves to umbrella a capture process of three dimensionally filming events, be they synthetic, real or hyper real. This term is technology agnostic and will not be dated by the means taken in any given application.

The advent of virtual worlds has given a new push to this concept since they allow the creation of real-time animation by using camera moves and avatar control techniques that are not possible by using traditional film-making methods Machinima Island.

References

- http://www.braintrustdv.com/essays/image-future.html
- Debevec, Paul (2006). "Virtual Cinematography: Relighting through Computation" [1] (PDF). *Computer* (IEEE) **39** (8): 57–65. doi:10.1109/MC.2006.285. ISSN 0018-9162. Archived from the original [2] on 2007-02-09. Retrieved 2006-11-06.
- Jhala, Arnav (2005) (PDF). *Introduction to Virtual Cinematography* [3]. North Carolina State University. Archived from the original [4] on 2007-01-02. Retrieved 2006-11-06.
- Newman, Bruce (2003). *'Matrix' sequel: Virtual Cinematography* [5]. San Jose Mercury News. Retrieved 2006-11-07.

Related

- Timeline of CGI in movies

References

[1] http://web.archive.org/web/20070209202337/http://www.debevec.org/Publications/DEBEVEC-IEEE-Computer-200608-high.pdf

[2] http://www.debevec.org/Publications/DEBEVEC-IEEE-Computer-200608-high.pdf

[3] http://web.archive.org/web/20070102004752/http://www.cs.unc.edu/~lastra/comp870/intro_vc,+Arnav+Jhala.pdf

[4] http://www.cs.unc.edu/~lastra/comp870/intro_vc,%20Arnav%20Jhala.pdf

[5] http://www.temple.edu/ispr/examples/ex03_05_13b.html

The Matrix

The Matrix	
 Theatrical release poster	
Directed by	• Andy Wachowski • Larry Wachowski
Produced by	Joel Silver
Written by	• Andy Wachowski • Larry Wachowski
Starring	• Keanu Reeves • Laurence Fishburne • Carrie-Anne Moss • Hugo Weaving • Joe Pantoliano
Music by	Don Davis
Cinematography	Bill Pope
Editing by	Zach Staenberg
Studio	• Village Roadshow Pictures • Silver Pictures
Distributed by	Warner Bros. Pictures
Release date(s)	• March 31, 1999
Running time	136 minutes
Country	United States
Language	English
Budget	$63 million
Box office	$463,517,383[1]

The Matrix is a 1999 American science fiction action film written and directed by Larry and Andy Wachowski. The film stars Keanu Reeves, Laurence Fishburne, Carrie-Anne Moss, Joe Pantoliano, and Hugo Weaving, and was first released in the United States on March 31, 1999. The success of the film led to the release of two feature film sequels, and the *Matrix* franchise was further expanded through the production of comic books, video games, and animated short films.

The film depicts a future in which reality as perceived by most humans is actually a simulated reality created by sentient machines to pacify and subdue the human population, while their bodies' heat and electrical activity are used as an energy source. Upon learning this, computer programmer "Neo" is drawn into a rebellion against the machines, involving other people who have been freed from the "dream world" and into reality.

The film contains many references to the cyberpunk and hacker subcultures; philosophical and religious ideas such as René Descartes' evil genius, the Allegory of the Cave, the brain in a vat thought experiment; and homages to *Alice's Adventures in Wonderland,* Hong Kong action cinema, spaghetti westerns, dystopian fiction, and Japanese animation.

Plot

Computer programmer Thomas Anderson (Keanu Reeves) is secretly a hacker known as "Neo". He is restless, and driven to learn the meaning of cryptic references to the "Matrix" appearing on his computer. A female hacker named Trinity (Carrie-Anne Moss) confirms that a man named Morpheus (Laurence Fishburne) can help him; however, three sinister Agents, led by Agent Smith (Hugo Weaving), arrest Neo and attempt to prevent him from collaborating with Morpheus. Undeterred, Neo meets with Morpheus and agrees to follow him by swallowing an offered red pill. Neo abruptly awakens in a liquid-filled vessel, connected along with thousands of other people to an elaborate electrical structure. He is rescued by Morpheus and brought aboard a levitating ship called the *Nebuchadnezzar*.

Morpheus tells Neo that the year is approximately 2199, and humans are fighting against intelligent machines that were created early in the 21st century and have since taken control of the Earth's surface. The machines harvest the bioelectrical energy of humans, who are kept docile within the Matrix, a simulated reality of the world as it was in 1999. Morpheus and his crew belong to a group of free humans who "unplug" others from the Matrix and recruit them to their rebellion against the Machines, and who are able to gain superhuman abilities within the Matrix by using their understanding of its true nature to manipulate its physical laws. Neo is warned that fatal injuries within the Matrix will also kill one's physical body, and that the Agents he previously met are powerful sentient computer programs that eliminate human threats to the system. Morpheus believes that Neo is "the One", a man prophesied to end the war.

The group enters the Matrix to visit the Oracle (Gloria Foster), who predicted the emergence of the One. The Oracle implies that Neo is not the One, but warns that Neo must soon choose between his life and that of Morpheus.

The group is ambushed by Agents and tactical police, leading to the death of the crew member Mouse (Matt Doran). Morpheus allows himself to be captured to let Neo and the crew escape. Their ally Cypher (Joe Pantoliano) had betrayed them, making a deal for a permanent return to the comfortable life of the Matrix in exchange for giving Morpheus to the Agents. Cypher murders three crew members before he is killed by Tank (Marcus Chong).

The Agents drug and interrogate Morpheus in an attempt to learn his access codes to the mainframe computer in Zion, the humans' subterranean refuge in the real world. Neo and Trinity return to the Matrix and rescue their leader; in the process, Neo becomes more confident in his ability to manipulate the Matrix, ultimately dodging bullets fired at him.

Morpheus and Trinity use a telephone to exit the Matrix, but Neo is ambushed by Agent Smith. He stands his ground and defeats Smith, but flees when the Agent possesses another body. Meanwhile, in the real world "sentinel" machines converge on the *Nebuchadnezzar*. Upon reaching another exit, Neo is shot and killed by Agent Smith. Trinity, who is standing over Neo in the real world, whispers that the Oracle told her she would fall in love with the One. She kisses Neo and restores his life. Neo revives with new power to perceive and control the Matrix, and effortlessly destroys Agent Smith, before returning to the real world in time for the ship's EMP weapon to destroy the attacking sentinels.

The film ends with Neo back in the Matrix, making a telephone call promising that he will demonstrate to the people imprisoned in the Matrix that "anything is possible." He hangs up the phone and flies into the sky.

Cast

- Keanu Reeves as Thomas A. Anderson/Neo: A computer programmer in Metacortex corporation who moonlights as a hacker. Recruited by Morpheus to fight the machines. Will Smith turned down the role of Neo to make *Wild Wild West*, because of skepticism over the film's ambitious bullet time special effects.[2] He later stated he was "not mature enough as an actor" at that time,[2] and that if given the role, he "would have messed it up".[3] [4] Nicolas Cage also turned down the role because of "family obligations".[5]
- Laurence Fishburne as Morpheus: A human freed from the Matrix, captain of the Nebuchadnezzar. He finds Neo and helps him learn the truth. Sean Connery declined the role of Morpheus.[6]
- Carrie-Anne Moss as Trinity: Freed by Morpheus, crewmember of the Nebuchadnezzar and Neo's romantic interest.
- Hugo Weaving as Agent Smith: A sentient "Agent" program of the Matrix whose purpose is to destroy Zion and stop humans from getting out of the Matrix. Unlike other agents, he has ambitions to free himself from his duties.
- Joe Pantoliano as Cypher: Another human freed by Morpheus, who betrays Morpheus to the Agents to ensure his return to the Matrix where he can be someone famous and live a life of luxury.
- Julian Arahanga as Apoc: A freed human and crew member on the Nebuchadnezzar.
- Anthony Ray Parker as Dozer: A "natural" human born outside of the Matrix, and pilot of the *Nebuchadnezzar*.
- Marcus Chong as Tank: the "operator" of the *Nebuchadnezzar*, he is Dozer's brother, and like him was born outside of the Matrix.
- Matt Doran as Mouse: A freed human and programmer on the Nebuchadnezzar.
- Gloria Foster as the Oracle: Exiled sentient computer program who still resides in the Matrix, helping the freed humans with her foresight and wisdom.
- Belinda McClory as Switch: A human freed by Morpheus and crew member of the Nebuchadnezzar.
- Paul Goddard as Agent Brown: One of two sentient "Agent" programs in the Matrix who work with Agent Smith to destroy Zion and stop humans escaping the system.
- Robert Taylor as Agent Jones: Second sentient "Agent" program working with Agent Smith.
- Ada Nicodemou as DuJour (The White Rabbit Girl), a reference to the White Rabbit in Alice's Adventures in Wonderland

Production

The Matrix was a co-production of Warner Bros. and Australian Village Roadshow Pictures, and all but a few scenes were filmed at Fox Studios in Sydney, Australia, and in the city itself. Recognizable landmarks were not included in order to maintain the setting of a generic American city.[7] Nevertheless, the Sydney Harbour Bridge, ANZAC Bridge, University of Technology, Sydney, AWA Tower, Martin Place and a Commonwealth Bank branch are visible in some shots, as is signage on buildings for the Sydney offices of MMI Insurance (now Allianz), AON, Citigroup, Telstra, Westpac, KMPG, Ernst & Young and IBM Corporation among others. Other clues to the filming location include left-hand traffic flow and signs featuring British English terminology and spellings such as "lift" and "authorised" (rather than the American English "elevator" and "authorized").

Subtle nods were included to Chicago, Illinois, the home city of the directors, through a subtly placed picture of the Chicago skyline, city maps, the destination of the subway train during the subway station fight between Neo and Agent Smith saying "Loop", and the names of streets: Wells, Lake, Franklin, Erie, State, Balbo, Wabash, Adams.

The rooftop set that Trinity uses to escape from Agent Jones early in the film was left over from the production of *Dark City*, which has been remarked upon due to the thematic similarities of the films.[8] According to *The Art of the Matrix*, at least one filmed scene and a variety of short pieces of action were omitted from the final cut, and have (to date) not been published.

The Wachowski Brothers were keen that all involved understood the thematic background of the movie. For example, the book used to conceal disks early in the movie, *Simulacra and Simulation*, a 1981 work by the French philosopher Jean Baudrillard, was required reading for most of the principal cast and crew.

Comic book artists Geof Darrow and Steve Skroce worked on *The Matrix* as concept and storyboard artists respectively.

Production design

In the film, the code that comprises the Matrix itself is frequently represented as downward-flowing green characters. This code includes mirror images of half-width kana characters and Western Latin letters and numerals. Generally, the film's production design placed a bias towards its distinctive green color for scenes set within the Matrix, whereas there is an emphasis on the color blue during the scenes set in the real world. In addition, grid-patterns were incorporated into the sets for scenes inside the Matrix, intended to convey the cold, logical and artificial nature of that environment.[9]

The "digital rain" is strongly reminiscent of similar computer code in the film *Ghost in the Shell*, an acknowledged influence on the *Matrix* series (see below). The color green reflects the green tint commonly used on early monochrome computer monitors.

Visual effects

> As for artistic inspiration for bullet time, I would credit Otomo Katsuhiro, who co-wrote and directed *Akira*, which definitely blew me away, along with director Michel Gondry. His music videos experimented with a different type of technique called view-morphing and it was just part of the beginning of uncovering the creative approaches toward using still cameras for special effects. Our technique was significantly different because we built it to move around objects that were themselves in motion, and we were also able to create slow-motion events that 'virtual cameras' could move around – rather than the static action in Gondry's music videos with limited camera moves.
>
> —John Gaeta[10]

Part of the lobby scene

The film is known for popularizing the use of a visual effect known as "bullet time", which allows the viewer to explore a moment progressing in slow-motion as the camera appears to orbit around the scene at normal speed.[11]

The method used for creating these effects involved a technically expanded version of an old art photography technique known as time-slice photography, in which a large number of cameras are placed around an object and triggered nearly simultaneously.[11] Each camera is a still-picture camera, and not a motion picture camera, and it contributes just one frame to the video sequence. When the sequence of shots is viewed as in a movie, the viewer sees what are in effect two-dimensional "slices" of a three-dimensional moment. Watching such a "time slice" movie is akin to the real-life experience of walking around a statue to see how it looks from different angles. The positioning of the still cameras can be varied along any desired smooth curve to produce a smooth looking camera motion in the finished clip, and the timing of each camera's firing may be delayed slightly,[11] so that a motion scene can be executed (albeit over a very short period of real time).

Some scenes in *The Matrix* feature the "time-slice" effect with completely frozen characters and objects. Film interpolation techniques improved the fluidity of the apparent "camera motion". The effect was further expanded upon by the Wachowski brothers and the visual effects supervisor John Gaeta so as to create "bullet time", which incorporates temporal motion, so that rather than being totally frozen the scene progresses in slow and variable motion.[11] Engineers at Manex Visual Effects pioneered 3-D visualization planning methods to move beyond mechanically fixed views towards more complicated camera paths and flexibly moving interest points. There is also an improved fluidity through the use of non-linear interpolation, digital compositing, and the introduction of computer generated "virtual" scenery. The movie was rendered on a FreeBSD cluster farm.[12]

The objective of the bullet time shots in *The Matrix* was to creatively illustrate "mind over matter" type events as captured by a "virtual camera". However, the original technical approach was physically bound to pre-determined perspectives, and the resulting effect only suggests the capabilities of a true virtual camera.

The evolution of photogrametric and image-based computer-generated background approaches in *The Matrix*'s bullet time shots set the stage for later innovations unveiled in the sequels *The Matrix Reloaded* and *The Matrix Revolutions*. Virtual Cinematography (CGI-rendered characters, locations, and events) and the high-definition "Universal Capture" process completely replaced the use of still camera arrays, thus more closely realizing the "virtual camera".

Music

The film's score was composed by Don Davis. He noted that mirrors appear frequently in the movie: reflections of the blue and red pills are seen in Morpheus's glasses; Neo's capture by Agents is viewed through the rear-view mirror of Trinity's Triumph Speed Triple motorcycle; Neo observes a broken mirror mending itself; reflections warp as a spoon is bent; the reflection of a helicopter is visible as it approaches a skyscraper. (The film also frequently references the book *Alice's Adventures in Wonderland*, which has a sequel entitled *Through the Looking-Glass*.) Davis focused on this theme of reflections when creating his score, alternating between sections of the orchestra and attempting to incorporate contrapuntal ideas.[13]

In addition to Davis' score, *The Matrix* soundtrack also features music from acts such as Rammstein, Rob Dougan, Rage Against the Machine, Propellerheads, Ministry, Deftones, Monster Magnet, The Prodigy, Rob Zombie, Meat Beat Manifesto, and Marilyn Manson. Other pieces from artists such as Duke Ellington, Django Reinhardt, and Massive Attack are included in the film, but not featured on the soundtrack.

Influences

The Matrix is arguably the ultimate cyberpunk artifact.

William Gibson, 2003-01-28[14]

The Matrix makes numerous references to recent films and literature, and to historical religions and philosophy. *The Matrix*'s premise resembles Plato's Allegory of the Cave, Calderon de la Barca's Life is a Dream, Edwin Abbott Abbott's *Flatland*, René Descartes's evil genius, Georges Gurdjieff's *The Sleeping Man*,[15] Kant's reflections on phenomena versus the Ding an sich, and the brain in a vat thought experiment.

Jean Baudrillard's *Simulacra and Simulation* is featured in the film, and was required reading for the actors.[16] However, Baudrillard commented that *The Matrix* misunderstands and distorts his work.[17] [18]

In Postmodern thought, interpretations of *The Matrix* often reference Baudrillard's philosophy to demonstrate that the movie is an allegory for contemporary experience in a heavily commercialized, media-driven society, especially of the developed countries. The influence of the matrixial theory of Bracha Ettinger articulated in a series of books and essays from the end of the 1980s onwards was brought to the public's attention through the writings of art historians such as Griselda Pollock[19] [20] and film theorists such as Heinz-Peter Schwerfel.[21]

There are similarities to several works by science fiction author Philip K. Dick,[22] [23] [24] as well as cyberpunk works such as *Neuromancer* by William Gibson.[25] The film's use of the term "Matrix" is adopted from Gibson's novel.[26]

Japanese director Mamoru Oshii's *Ghost in the Shell* was a strong influence. Producer Joel Silver has stated that the Wachowski brothers first described their intentions for *The Matrix* by showing him that anime and saying, "We wanna do that for real".[27] [28] Mitsuhisa Ishikawa of Production I.G, which produced *Ghost in the Shell*, noted that the anime's high-quality visuals were a strong source of inspiration for the Wachowski brothers. He also commented, "... cyberpunk films are very difficult to describe to a third person. I'd imagine that *The Matrix* is the kind of film that was very difficult to draw up a written proposal for to take to film studios". He stated that since *Ghost in the Shell* had gained recognition in America, the Wachowski brothers used it as a "promotional tool".[29] Stanley Kubrick's *2001: A Space Odyssey* was another science fiction film that helped inspire the visual style of *The Matrix*.[30]

Reviewers have commented on similarities between *The Matrix* and other late-1990s films such as *Strange Days*, *Dark City*, and *The Truman Show*.[31] [32] [33] Comparisons have also been made to Grant Morrison's comic series *The Invisibles*; Morrison believes that the Wachowski brothers essentially plagiarized his work to create the film.[34] [35] Comparisons have also been made between *The Matrix* and the books of Carlos Castaneda.[36] In addition, the similarity of the film's central concept to a device in the long running series *Doctor Who* has also been noted. As in the film, the Matrix of that series (introduced in the 1976 serial *The Deadly Assassin*) is a massive computer system which one enters using a device connecting to the head, allowing users to see representations of the real world and change its laws of physics; but if killed there, they will die in reality.[37]

Release

The Matrix was first released on March 31, 1999. It earned $171 million in North America and over $292 million in foreign box offices, for a total of $463 million worldwide,[1] and later became the first DVD to sell more than three million copies in the US.[38] The Ultimate Matrix Collection was released on HD DVD on May 22, 2007[39] and on Blu-ray on October 14, 2008.[40]

The film was also released standalone in a 10th anniversary edition Blu-ray in the Digibook format on March 31, 2009, 10 years to the day after the movie was released theatrically.[41]

Reception

The Matrix received positive reviews from most critics,[42] and is widely regarded as one of the greatest science fiction films of all time.[43] [44] [45] On Metacritic, it has a consensus forming that it presented an "ingenious" blend of Hong Kong action cinema, innovative visual effects and an imaginative vision.[46] *Rotten Tomatoes* reported that 87% of critics gave the film positive reviews, with an average score of 7.4/10, based upon a sample of 126 reviews.[46] The site reported that 68% of selected notable critics gave the film a positive review, based upon a sample of 28.[47] At *Metacritic*, which assigns an average rating out of 100 to reviews from mainstream critics, the film received an average score of 73 upon its DVD release, based on 35 reviews.[42]

Philip Strick commented in *Sight & Sound*, "if the Wachowskis claim no originality of message, they are startling innovators of method," praising the film's details and its "broadside of astonishing images".[48] Roger Ebert praised the film's visuals and premise, but disliked the third act's focus on action.[31] Similarly, *Time Out* praised the "entertainingly ingenious" switches between different realities, Hugo Weaving's "engagingly odd" performance, and the film's cinematography and production design, but concluded, "the promising premise is steadily wasted as the film turns into a fairly routine action pic ... yet another slice of overlong, high concept hokum."[49] Other reviewers criticised the relative humorlessness and self-indulgence of the movie.[50] [51] [52]

In 2001, *The Matrix* was placed 66th in the American Film Institute's "100 Years... 100 Thrills" list. In 2007, *Entertainment Weekly* called *The Matrix* the best science-fiction piece of media for the past 25 years.[53] The film is also ranked number 39 on Empire's "The 500 Greatest Movies of All Time."[54] *The Matrix* was voted as the fourth best sci-fi film in the 2011 list *Best in Film: The Greatest Movies of Our Time*, based on a poll conducted by ABC and *People*.

Several science fiction creators commented on the film. Author William Gibson, a key figure in cyberpunk fiction, called the film "an innocent delight I hadn't felt in a long time," and stated, "Neo is my favourite-ever science fiction hero, absolutely."[55] Joss Whedon called the film "my number one" and praised its storytelling, structure and depth, concluding, "It works on whatever level you want to bring to it."[56] Filmmaker Darren Aronofsky commented, "I walked out of *The Matrix* ... and I was thinking, 'What kind of science fiction movie can people make now?' The Wachowskis basically took all the great sci-fi ideas of the 20th century and rolled them into a delicious pop culture sandwich that everyone on the planet devoured."[57] Director M. Night Shyamalan praised the Wachowskis' passion for the film, saying, "Whatever you think of *The Matrix*, every shot is there because of the passion they have! You can see they argued it out!"[58]

Awards and nominations

The Matrix received Oscars for film editing, sound effects editing, visual effects, and sound. The filmmakers were competing against other films with established franchises, like *Star Wars Episode I: The Phantom Menace*, yet they managed to sweep all four of their nominations.[59] [60] In 1999, it won Saturn Awards for **Best Science Fiction Film** and **Best Direction**.[61] *The Matrix* also received BAFTA awards for **Best Sound** and **Best Achievement in Special Visual Effects**, in addition to nominations in the cinematography, production design and editing categories.[62]

Award	Category	Name	Outcome
72nd Academy Awards	Film Editing	Zach Staenberg	Won
	Sound Mixing	John T. Reitz, Gregg Rudloff, David E. Campbell, David Lee	Won
	Sound Editing	Dane A. Davis	Won
	Visual Effects	John Gaeta	Won

American Film Institute Lists

- AFI's 100 Years...100 Thrills – #66
- AFI's 100 Years...100 Heroes and Villains:
  - Neo (Thomas Anderson) – Nominated Hero
  - Agent Smith – Nominated Villain
- AFI's 100 Years...100 Movies (10th Anniversary Edition) – Nominated
- AFI's 10 Top 10 – Nominated Science Fiction Film

Legacy

The Matrix has had a strong effect on action film-making in Hollywood. It set a new standard for cinematic fight scenes[63] by hiring acclaimed choreographers (such as Yuen Woo-ping) from the Hong Kong action cinema scene, well known for its production of martial arts films. The success of *The Matrix* put those choreographers and their techniques in high demand by other filmmakers who wanted fights of similar sophistication: for example, wire work was employed in *X-Men* (2000),[63] and Yuen Woo-ping's brother Yuen Cheung-Yan was choreographer on *Daredevil* (2003).

Following *The Matrix*, films made abundant use of slow-motion, spinning cameras, and, often, the bullet time effect of a character freezing or slowing down and the camera dollying around them. The ability to slow down time enough to distinguish the motion of bullets was used as a central gameplay mechanic of several video games, including *Max Payne*, in which the feature was explicitly referred to as "bullet time" (although the game went into production before the film was released). *The Matrix*'s signature special effect has also been parodied numerous times, in comedy films such as *Scary Movie*, *Deuce Bigalow: Male Gigolo*, *Shrek* and *Kung Pow! Enter the Fist*; in animated TV series such as *The Simpsons*, *Fairly Oddparents* and *Family Guy*; in the OVA series *FLCL*; and in video games such as *Conker's Bad Fur Day*, along with a more elaborate parody called *Marx Reloaded*, in which the central relationship between Neo and Morpheus is represented as an imaginary encounter between Karl Marx and Leon Trotsky.

Franchise

The film's mainstream success led to the making of two sequels, *The Matrix Reloaded* and *The Matrix Revolutions*. These were filmed simultaneously during one shoot and released in two parts in 2003. The first film's introductory tale is succeeded by the story of the impending attack of the human enclave of Zion by a vast machine army. Neo also learns more about the history of the Matrix, his role as the One and the prophecy that he will end the war. The sequels also incorporate longer and more ambitious action scenes, as well as improvements in bullet time and other visual effects.

Also released was *The Animatrix*, a collection of nine animated short films, many of which were created in the same Japanese animation style that was a strong influence on the live trilogy. *The Animatrix* was overseen and approved by the Wachowski brothers who only wrote four of the segments themselves and did not direct any of them; much of the project was developed by notable figures from the world of anime. Four of the films were originally released on the series' official website [64]; one was shown in cinemas with the Warner Bros. movie *Dreamcatcher*; the others first appeared with the DVD release of all nine shorts. Several of the films were shown first on UK television prior to their DVD release.

The franchise also contains three video games: *Enter the Matrix* (2003), which contains footage shot specifically for the game and chronicles events taking place before and during *The Matrix Reloaded*; *The Matrix Online* (2004), a MMORPG which continued the story beyond *The Matrix Revolutions*; and *The Matrix: Path of Neo* (2005), which focuses on situations based on Neo's journey through the trilogy of films.

The Matrix Comics is a series of comics and short stories set in the world of *The Matrix*, written and illustrated by figures from the comics industry. Most of the comics were originally presented for free on the official *Matrix* website;[65] they were later republished, along with some new material, in two printed trade paperback volumes.

See also

- Artificial intelligence
- Brain in a vat
- Henosis
- Simulated reality

References

Notes

[1] Box Office Mojo: The Matrix (http://www.boxofficemojo.com/movies/?id=matrix.htm). URL retrieved 24 June 2009.

[2] Lawrence, Will (February 2007). "The Empire Interview: In conversation with Will Smith". *Empire* (EMAP) (212): 109. "Honestly, I didn't think they could do it, it was too ambitious. I saw *Bound* and I loved it. *The Matrix* is exactly what they pitched, but they were designing those cameras to get those freeze-frames, and I was like, "If that doesn't work, the movie looks ridiculous." I didn't feel comfortable with the level of importance placed on that effect working properly. ... That's probably the only one that I turned down that I shouldn't have, but when you see somebody do it like Keanu you think, "Thank God." I don't think I was mature enough as an actor at that point to get out of the way and just let it be and allow the directors to make the movie. I would have been trying to make jokes. Now I would have loved to take a shot and see what I would have done with it and I know now I could absolutely have been mature enough to get out the way. But back then I don't think I was."

[3] Hillner, Jennifer. "I, Robocop" (http://www.wired.com/wired/archive/12.07/smith_pr.html). *Wired*. Condé Nast Publications. .

[4] Riggs, Ransom. " 5 million-dollar mistakes by movie stars (http://www.cnn.com/2008/LIVING/worklife/10/20/mf.rejected.movies/index.html?iref=mpstoryview)." *CNN*. Retrieved October 20, 2008.

[5] Larry Carroll (2007-12-07). "Will Smith Snagged 'I Am Legend' From Schwarzenegger, But Can You Imagine Nicolas Cage In 'The Matrix'?" (http://www.mtv.com/movies/news/articles/1576048/20071207/story.jhtml). MTV. . Retrieved 2007-12-08.

[6] The League of Extraordinary Gentlemen: DVD Behind The Scenes Footage

[7] Behind-the-scenes documentary "HBO First Look: Making the Matrix"

[8] Ebert, Roger (November 6, 2005). "Great Movies: Dark City" (http://rogerebert.suntimes.com/apps/pbcs.dll/article?AID=/20051106/REVIEWS08/511060302/1023). *Chicago Sun-Times*. . Retrieved December 18, 2006.

[9] Costume designer Kym Barrett, production designer Owen Paterson and cinematographer Bill Pope, interviewed in *The Matrix Revisited* (Chapter 7).

[10] "200 Things That Rocked Our World: Bullet Time". *Empire* (EMAP) (200): 136. February 2006.

[11] Green, Dave (1999-06-05). "Better than SFX" (http://www.guardian.co.uk/film/1999/jun/05/features2). *The Guardian* (London). . Retrieved December 18, 2009.

[12] "Comment about the use of FreeBSD (5:50)" (http://www.youtube.com/watch?v=bAsYz5pVwyc#t=5m45s). Youtube.com. 2008-01-23. . Retrieved 2012-01-29.

[13] Don Davis, interviewed in *The Matrix Revisited* (Chapter 28). A transcript of his comments may be found online: (http://web.archive.org/web/20070429165147/http://www.geocities.com/dondavismatrixnl/Dvdfeaturesdavis.html)

[14] "THE MATRIX: FAIR COP" (http://www.williamgibsonbooks.com/archive/2003_01_28_archive.asp#90244012), *The William Gibson Blog*

[15] Oksanen, Reijjo. "Planvan N. Go Interview" (http://www.katinkahesselink.net/sufi/zen.html). The Gurdjieff Internet Guide. (Retrieved 09–03–17).

[16] Post Jobs (2007-03-14). "Remember Baudrillard" (http://www.insidehighered.com/views/mclemee/mclemee135). Inside Higher Ed. . Retrieved 2012-01-29.

[17] "Le Nouvel Observateur with Baudrillard" (http://web.archive.org/web/20080113012028/http://www.empyree.org/divers/Matrix-Baudrillard_english.html). Le Nouvel Observateur. Archived from the original (http://www.empyree.org/divers/Matrix-Baudrillard_english.html) on 2008-01-13. . Retrieved 2010-01-31.

[18] "Warner Bros" (http://whatisthematrix.warnerbros.com/rl_cmp/new_phil_fr_hanley2.html). Whatisthematrix.warnerbros.com. . Retrieved 2012-01-29.

[19] Griselda Pollock, "Does Art Think?" In: Dana Arnold and Margaret Iverson (eds.) *Art and Thought*. Oxford: Basil Blackwell, 2003. ISBN 0-631-22715-6

[20] Griselda Pollock, "Inscritions in the Feminine" In: Catherine de Zegher (eds), *Inside the Visible*. MIT Press, 1996

[21] Heinz-Peter Schwerfel, *Kino and Kunst*, Koln: Dumont, 2003.

[22] Rose, Frank. "The Second Coming of Philip K. Dick" (http://www.wired.com/wired/archive/11.12/philip_pr.html). Wired magazine. .

[23] Zenko, Darren (2007-04-29). "Not another Philip K. Dick movie" (http://www.thestar.com/entertainment/article/208471). The Toronto Star. . Retrieved 2010-05-25.

[24] Axmaker, Sean (2002-06-25). "Philip K. Dick's dark dreams still fodder for films" (http://www.seattlepi.com/movies/76011_dick26.shtml). Seattle Post Intelligencer. .

[25] "The Matrix: Fair Cop" (http://www.williamgibsonbooks.com/blog/2003_01_01_archive.asp#90244012). URL retrieved 7 July 2006.

[26] Leiren-Young, Mark (6 January 2012). "Is William Gibson's 'Neuromancer' the Future of Movies?" (http://thetyee.ca/ArtsAndCulture/2012/01/06/Neuromancer-Movie/). *The Tyee*. . Retrieved 16 January 2012. "One of the obstacles in the selling of this movie to the industry at large is that everyone says, 'Oh, well, *The Matrix* did it already.' Because *The Matrix* – the very word 'matrix' – is taken from *Neuromancer*, they stole that word, I can't use it in our movie."

[27] Joel Silver, interviewed in "Scrolls to Screen: A Brief History of Anime" featurette on *The Animatrix* DVD.

[28] Joel Silver, interviewed in "Making *The Matrix*" featurette on *The Matrix* DVD.

[29] Mitsuhisa Ishikawa, interviewed in *The South Bank Show*, episode broadcast 19 February 2006 (http://www.itv.com/page.asp?partid=2701)

[30] Ebert, Roger. "The Wachowskis: From "2001" to "The Godfather" to "The Matrix"" (http://rogerebert.suntimes.com/apps/pbcs.dll/article?AID=/20081002/COMMENTARY/810020297). *Chicago Sun-Times*. . Retrieved 2010-01-30.

[31] Roger Ebert's review of *The Matrix* (http://rogerebert.suntimes.com/apps/pbcs.dll/article?AID=/19990331/REVIEWS/903310303/1023). URL retrieved 21 August 2006.

[32] "The Matrix (1999) – Channel 4 Film review" (http://www.channel4.com/film/reviews/film.jsp?id=105863). URL retrieved 21 August 2006.

[33] Rowley, Stephen (2003-06-18). "What Was the Matrix?" (http://www.sterow.com/?p=229). *sterow.com*. . Retrieved 9 January 2012.

[34] "Poor Mojo Newswire: Suicide Girls Interview with Grant Morrison (http://www.poormojo.org/pmjadaily/archives/002657.html)". URL retrieved 31 July 2006.

[35] Morrison, Grant (2011). *Supergods: Our World in the Age of the Superhero*. London: Random House/Jonathan Cape. p. 315. "I was taken to see *The Matrix* ... and saw what seemed to me my own combination of ideas enacted on the screen: fetish clothes, bald heads, kung fu, and magic, witnessing the Gnostic invasion of the Hollywood mainstream."

[36] "Matrix and Carlos Castaneda" (http://www.consciencia.org/castaneda/castaneda-matrix.html). Consciencia.org. 2006-12-04. . Retrieved 2012-01-29.

[37] Condon, Paul. *The Matrix Unlocked*. 2003. Contender. pp.141–3. ISBN 1-84357-093-9

[38] *"Press release - August 1, 2000 - The Matrix DVD: The first to sell 3 million"* (http://web.archive.org/web/20080305104849/http://whatisthematrix.warnerbros.com/rl_cmp/rl_press_August_01_00.html). *Whatisthematrix.warnerbros.com. Burbank: Warner Bros., Inc.. August 1, 2000. Archived from the original* (http://whatisthematrix.warnerbros.com/rl_cmp/rl_press_August_01_00.html) *on March 5, 2008. . Retrieved 10 April 2012.*

[39] Warner Home Video (2007-03-23). "The Matrix is Coming to HD DVD" (http://www.comingsoon.net/news/movienews.php?id=19508). Comingsoon.net. . Retrieved 2007-03-23.

[40] Warner Home Video (2008-07-25). "'Ultimate Matrix' Blu-ray Coming in October" (http://www.highdefdigest.com/news/show/Disc_Announcements/Warner/Ultimate_Matrix_Blu-ray_Coming_in_October/1944). highdefdigest.com. . Retrieved 2008-08-18.

[41] "Warner Home Video sends over details on a 10th Anniversary Blu-ray release" (http://www.dvdactive.com/news/releases/the-matrix.html). Dvdactive.com. . Retrieved 2009-12-13.

[42] "The Matrix (1999): Reviews" (http://www.metacritic.com/video/titles/matrix). *Metacritic*. CNET Networks, Inc. . Retrieved 2008-07-11.

[43] Heritage, Stuart (21 October 2010). "The Matrix: No 13 best sci-fi and fantasy film of all time" (http://www.guardian.co.uk/film/2010/oct/21/matrix-wachowskis-science-fiction). *Guardian* (London). .

[44] "Top 25 Sci-Fi Movies of All Time – Movies Feature at IGN" (http://movies.ign.com/articles/677/677739p4.html). Movies.ign.com. . Retrieved 2012-01-29.

[45] "Online Film Critics Society: OFCS Top 100: Top 100 Sci-Fi Films" (http://www.ofcs.org/2010/09/ofcs-top-100-top-100-sci-fi-films.html). Ofcs.org. 2010-09-24. . Retrieved 2012-01-29.

[46] "The Matrix Movie Reviews" (http://www.rottentomatoes.com/m/matrix/). *Rotten Tomatoes*. IGN Entertainment, Inc. . Retrieved 2008-07-11.

[47] "The Matrix Movie Reviews, Top Critics" (http://www.rottentomatoes.com/m/matrix/?critic=creamcrop). *Rotten Tomatoes*. IGN Entertainment, Inc. . Retrieved 2008-07-11.

[48] "*Sight & Sound* review of *The Matrix*" (http://www.bfi.org.uk/sightandsound/review/151/). . Retrieved 2007-02-03.

[49] "Time Out Film Review – The Matrix" (http://www.timeout.com/film/reviews/72947/the-matrix.html). *Time Out Film Guide*. Time Out. . Retrieved 2012-01-16.

[50] "Critical review of The Matrix" (http://onfilm.chicagoreader.com/movies/capsules/17448_MATRIX.html). . Retrieved 2007-02-03.

[51] "Negative review of The Matrix" (http://www.salon.com/ent/movies/reviews/1999/04/02reviewa.html). . Retrieved 2007-02-03.

[52] "Extremely negative review of The Matrix" (http://www.nicksflickpicks.com/matrix.html). . Retrieved 2008-05-09.

[53] Jeff Jensen (2007-05-07). "The Sci-Fi 25" (http://www.ew.com/ew/article/0,,20036782_20037403_20037541_25,00.html). Entertainment Weekly. . Retrieved 2007-05-07.

[54] "Empire Features" (http://www.empireonline.com/500/89.asp). Empireonline.com. . Retrieved 2009-12-13.

[55] *The Art of the Matrix*, p.451

[56] "The 201 Greatest Movies of all Time". Empire (Issue 201). March 2006. p. 98.

[57] Darren Aronofsky, quoted in the article " The Outsider (http://www.wired.com/wired/archive/14.11/outsider.html)", *Wired*. November 2006 issue (pp. 224)

[58] M. Night Shyamalan, quoted in the article 'OSCAR FILMS/FIRST TIMERS; A Director With a Sense of Where He's Going' " (http://query.nytimes.com/gst/fullpage.html?res=9F05E3DE1038F931A25750C0A9669C8B63&sec=&spon=&pagewanted=3)", *New York Times*. March 12, 2000

[59] "The 72nd Academy Awards (2000) Nominees and Winners" (http://www.oscars.org/awards/academyawards/legacy/ceremony/72nd-winners.html). *oscars.org*. . Retrieved 2011-11-19.

[60] "The Wachowski Brothers" (http://www.tribute.ca/directors/bios/10799.htm). Tribute magazine. . Retrieved 2006-12-31.

[61] "Saturn Awards" (http://www.saturnawards.org/past.html#film). SaturnAwards.org. . Retrieved 2006-12-31.

[62] "BAFTA Film Winners 1990–1999" (http://www.bafta.org/site/webdav/site/myjahiasite/shared/import/Film_Winners_1990-1999.pdf) (PDF). BAFTA.org. . Retrieved 2006-12-31.

[63] Jeff Jensen (2000-07-21). "Generating *X*" (http://www.ew.com/ew/article/0,,276737,00.html). *Entertainment Weekly*. . Retrieved 2008-12-31. "There was also debate over the style of the film's fight sequences, thanks to the new standard set by The Matrix, which hit while X-Men was in preproduction. Hence, the movie features some high-flying Matrix-y martial-arts choreography by Corey Yuen (Romeo Must Die)."

[64] http://www.whatisthematrix.com

[65] "Comics" (http://web.archive.org/web/20070815230811/http://whatisthematrix.warnerbros.com/rl_cmp/comics_new_front.html). *whatisthematrix.warnerbros.com*. Archived from the original (http://whatisthematrix.warnerbros.com/rl_cmp/comics_new_front.html) on 15 August 2007. . Retrieved 11 April 2012.

Bibliography

- Meinhold, Roman (2009.). *Being in the Matrix: An Example of Cinematic Education in Philosophy*. Prajna Vihara. Journal of Philosophy and Religion. Bangkok, Assumption University. Vol.10., No.1–2,. pp. 235–252 ISSN 1513–6442, http://www.roman-meinhold.com/matrix.pdf.
- Spencer Lamm (editor); Larry and Andy Wachowski, Steve Skroce, Geof Darrow, Tani Kunitake, Warren Manser, Colin Grant, Zach Staenberg, Phil Oesterhouse, William Gibson (2000). *The Art of the Matrix*. Titan. p. 488. ISBN 1-84023-173-4.
- Josh Oreck (Director) (2001). *The Matrix Revisited* (DVD). Warner Bros.

Further reading

- Faller, Stephen (2004). *Beyond The matrix: revolutions and revelations* (http://books.google.ca/books?id=qOfhfi0ka0sC&lpg=PP1&dq=The Matrix&pg=PP1#v=onepage&q&f=true). Chalice Press. ISBN 0827202350
- Kapel, Matthew; William G. Doty (2006). *Jacking in to the Matrix Franchise: Cultural Reception And Inerpretation* (http://books.google.ca/books?id=a0AmxKeakMoC&lpg=PR1&dq=The Matrix&pg=PR1#v=onepage&q&f=true). Continuum. ISBN 0826419097
- Herbrechter, Stefan (2006). *The Matrix in theory* (http://books.google.ca/books?id=f7GcmwUl7lgC&lpg=PP1&dq=The Matrix&pg=PP1#v=onepage&q&f=true). Rodopi. ISBN 9042016396
- Irwin, William (2005). *The matrix and philosophy: welcome to the desert of the real* (http://books.google.ca/books?id=GSE5qlwGzCAC&lpg=PP1&dq=The Matrix&pg=PP1#v=onepage&q&f=true). Open Court. ISBN 0812695011

External links

- *The Matrix* (http://www.imdb.com/title/tt0133093/) at the Internet Movie Database
- *The Matrix* (http://www.allrovi.com/movies/movie/v177524) at AllRovi
- *The Matrix* (http://www.boxofficemojo.com/movies/?id=matrix.htm) at Box Office Mojo
- *The Matrix* (http://www.rottentomatoes.com/m/matrix/) at Rotten Tomatoes

Still_camera

A **still camera** is a type of camera used to take photographs. Traditional cameras capture light onto photographic film. Digital cameras use electronics, usually a charge coupled device (CCD) to store digital images in computer memory inside the camera. The digital image can be transferred to a computer for later processing.

The still camera may be divided in a number of types (note that some cameras fall into more than one category):

- Autofocus camera
- Box camera
- Compact camera
- Digital camera
- Disposable camera
- Field camera
- Folding camera
- Instant camera
- Live-preview digital camera
- Pinhole camera
- Press camera
- Rangefinder camera
- Single-lens reflex camera
- Stereo camera
- Twin-lens reflex camera
- View camera
- Viewfinder camera

See also

- Photography
- PictBridge

Mature_technology

A **mature technology** is a technology that has been in use for long enough that most of its initial faults and inherent problems have been removed or reduced by further development. In some contexts, it may also refer to technology which has not seen widespread use, but whose scientific background is well understood.[1]

One of the key indicators of a mature technology is the ease of use for both non-experts and professionals. Another indicator is a reduction in the rate of new breakthrough advances related to it—whereas inventions related to a (popular) immature technology are usually rapid and diverse,[2] and may change the whole use paradigm—advances to a mature technology are usually incremental improvements only.

Examples

Mature technologies

- Farming, most advances are in slight improvements of breeds or in pest reduction
- Motor vehicle, widely used by non-experts, the general principles have not changed for decades
- Telephone, though considered mature, mobile phones showed a rare potential for substantial changes even in such technologies
- Firearm, typified by assault rifle technology, most advances are slight improvements as manufacturers alter balances between weight, firepower, range, and accuracy

Technologies not yet fully mature

- Internet, with still partly conflicting technological standards[3]
- Computers (becoming more mature due to advances in user-friendly operating systems)
- Economic models (still showing high failure rates in economic prediction)

Immature technologies

- Nanotechnology, actual industrial applications limited so far
- Manned space flight, limited, failure-prone and uneconomic
- Quantum computers, so far mostly a theoretical concept
- Nuclear fusion power, mainly theoretical in practice as containment energy expenditure thus far outweighs yielded energy

See also

- Business cycle
- Technology lifecycle
- Technology readiness level

References

[1] Defining mature technology (in the sustainability and security of energy supply and utilization) (http://engineering.dartmouth.edu/rbaef/documents/mature.pdf) (from a RBAEF memo, Dartmouth College, Tuesday 04 December 2003)

[2] How technologies evolve (http://www.oecd.org/dataoecd/23/52/2956490.pdf) (from an OECD and IEA information paper, 2003)

[3] The Status and Importance of Web Services (http://www.registrysolutions.co.uk/sdmxDemo/notes/SDMXSiteWebServices.htm) (from the registrysolutions.co.uk website)

Article Sources and Contributors

Bullet_time *Source*: http://en.wikipedia.org/w/index.php?title=Bullet_time *Contributors*: 16@r, 1ForTheMoney, Aaronh, Adkliphuis, Adoniscik, Akeron, Alai, Alexforcefive, Alexius08, Alfakim, Ambil, Anville, Anyep, Apostrophe, Asasa64, Ataricodfish, AtaruMoroboshi, BGyss, Beardo, Beoba, Bevo, Blocky, BlooWilt, BlueAzure, Bob f it, Boneho Chane, BonehoChane, Booklover323, BrokenSegue, BrotherEstapol, Bse3, Chester Markel, Chris the speller, Commander Keane, Crzrussian, CyberSkull, Dale Arnett, Danlev, Danno uk, Dark mercy, Darkanius, Darkspartan4121, Darth Panda, Darthjarek, Dave-ros, Daverocks, DavidForthoffer, Davidhorman, Deltabeignet, Dennisjiewenliu, Dominicos, Dust Filter, DzinX, EEMIV, EboMike, Edward Z. Yang, Eekerz, Eliazar, Empty Buffer, Emurphy42, EngineerScotty, Epolk, FF2010, Falcon9x5, Fama Clamosa, Fastfission, Fidofetch, Figaro, FlashpointDFW, Fletcherism, Frankyboy5, Froodrook, GeeJo, Gentlemen, George100, Gfxguy, Gilliam, Girolamo Savonarola, Grandpafootsoldier, Gs68, Gscshoyru, Gurch, Guroadrunner, Hahnchen, Hbdragon88, Hmains, HoraceInkling, Hudd, Hyacinth, Hydrargyrum, I do not exist, Iaoth, Ihavenolife, Immblueversion, Infamous JC, Intranetusa, J.J.Sagnella, Jagged 85, Jason One, Jaz246, Jeffro77, Jesse Hannah, JiMidnite, Jleedev, JoeLoeb, John gaeta, Johnny Orcface, JohnnyZu, Jonathan Grynspan, Josephhjones, Joy, JukeJohn, Jwinius, Karaboom, Kegon, Kevathens, Kicking222, King of the Dancehall, Kmactane, Kmg90, Knepflerle, Koavf, Koyaanis Qatsi, Kyz, Lamro, Lee M, Leemsy, Legoboy2000, Liftarn, Litefantastic, Litis, Liyang, Lucuas, M0rphzone, MPaquette, MStraw, Mackeriv, Maikeru Go, Marblespire, Markplume, Mars Marine, Master of Puppets, MattSutton1, Max314, McGeddon, Meco, MegX, Metallurgist, Mgiganteus1, Minesweeper, Mlubrecht, Momus, Monk127, Monkeyman, Moshe Constantine Hassan Al-Silverburg, NAG3, NAveryW, NMarkRoberts, Nahald, Nicapicella, Nick R, Nihiltres, Nikola Smolenski, NorrYtt, OldsVistaCruiser, Oliver Pereira, Omphaloscope, Orca1 9904, Osman-pasha, Ost316, Ourai, Patriarch, Peter S., Phil Boswell, Philwelch, Pictureuploader, Pikawil, Pimlottc, Pmj, Populus, Professorburntguy, ProhibitOnions, Purplefeltangel, Qutezuce, Raistlin8r, RedEyesMetal, Reedbeta, Reify-tech, RevRagnarok, Richard D. LeCour, Riki, Riotgear, Rjwilmsi, Roadrunner, Rsugden, Ryanlintelman, Ryulong, SKA-ed for Life, Scientizzle, Sheldrake, Shj95, SimonP, Sir Nicholas de Mimsy-Porpington, Slowtiger, Slysplace, Sockatume, Squiddle, Srleffler, Ste1n, StephanCom, Stillnotelf, Stormwatch, Straterp, Stylinnprofilin, Supapuerco, Surfer smith 00, Synthe, TMC1982, TRobertson47, TaintedMustard, Terrypin, Thingg, Thiseye, Toypaj, Tregoweth, Tronno, UVnet, Undeuxtroiskid, Vendettax, Verbose, Vicarious, ViperSnake151, Vroman, Vsync, Weedle McHairybug, Wessmaniac, William Avery, WilliamSommerwerck, Woc2006, Wonchop, Wwagner, Wwwwolf, X201, Y2kcrazyjoker4, Yath, Ylbissop, 405 anonymous edits

Simulation *Source*: http://en.wikipedia.org/w/index.php?title=Simulation *Contributors*: 05proedl, 16@r, 6024kingedward1, A. Parrot, AManWithNoPlan, Aaron Solomon Adelman, Aavviof, Acdoyle2000, Addshore, Adriatikus, Aeroman213, Aerotone, Aetheling, Agilemolecule, Agostinobruzzone, Ahanberg, Alan Fitzgerald, Albert ip, Aleenf1, Amorymeltzer, Andre Engels, Andrevruas, AndrewDressel, Anomalocaris, Ap, Arasut, Arthena, Arthur Rubin, Asparagus, Atlant, Axlq, B7582, BIONICLE233, BMF81, BackwardsBoy, Bapho, BarryList, Barticus88, Bdalgarno, Bdel63, Behaafarid, Bel.Alena, Ben Ben, Bob Murray, Bobbajobb, Bobblehead, Bobet, Brucevdk, C.Fred, CLandsberg85, COMPFUNK2, Caitlin Timms, Capricorn42, Cddejong, Celticsfan730, ChemGardener, Chowbok, Cleared as filed, Closedmouth, Commander Keane, Cortes IDS5717C, Cyclonenim, DVdm, Dannyc77, Dastly75, Davehi1, David Wright, Davidcofer73, Dayorkmd, DeepAnim1, Dgibson, Dhchilies, Dirkbb, Dlacombe, Drbreznjev, Duk, Duoduoduo, Enoch the red, EnumaElish, Ephebi, Erkan Yilmaz, Eterry, Eve Hall, Ever.anon, EyalBrill, EyeSerene, Fabricationary, FactChecker1199, Felonyboy, Fenice, Fordmadoxfraud, Foxj, Franciscoesquembre, Frankenpuppy, Free Bear, Fullyloadedcamel, G716, GJeffery, Gaius Cornelius, Gbgsimulationjon, Gdawgrancid, Gdlong, Gene Hobbs, Geniac, Gfoz, Gh5046, Gianfranco, Giftlite, GraemeL, Graham87, Greensburger, Guy.other, Guyd, Gz33, Hadal, Haffner, Hesperian, Hodsondd, Holdendp, Hongooi, Ian Geoffrey Kennedy, Icairns, Icmauri84, Impact red, Indon, Inwind, J.delanoy, J04n, JamesLee, Jbergste, Jdu, JeremyA, Jeremykemp, Jeroen12321, Jhadley6, Jillglen, Jimbojimbo2, Jlewis501, Jm0130, Jmkim dot com, Joewski, Johnpseudo, Jonathan Drain, Jonemerson, Joneskarenl, Jonkerz, Jorbs3210, Joyous!, Jpk, Jskiles1, Juansempere, Jvs.cz, KJS77, Kahmed198, Kai.velten, Kak Dela?, Kayau, Kdakin, Kkorpinen, Kqharvey, KrakatoaKatie, Kubanczyk, Kuffner, Kuru, Kylin.Ma, LA2, LeavesFX, Leuko, Levineps, Lexor, Liao, LilHelpa, Lim Sem Chow, Lindsay658, Listowy, Llacy, Loopblair, Lorna24, Lupo, M7, MC10, MKFI, MacGyverMagic, Malcolmxl5, Manop, Marasmusine, Martarius, Master on EN, Maurice Carbonaro, Mausy5043, Maxnegro, McGeddon, Mdd, Mitaphane, Mlewis000, MrOllie, Mynocturama, Nargoth, Neparis, Nick Number, Nivix, NoNameXII, Normsch, Nv8200p, Olaf Davis, Oli Filth, Omicronpersei8, Ovenkat, Oxymoron83, Oğuz Ergin, PTJoshua, Petedarnell, Petrov pv, Philip Trueman, PhilipMW, Pixie2000, Poiuyt Man, Ponyo, Powo, Pranav v, Prolog, Pwaddell, R'n'B, R. fiend, RDBrown, RGeurtz, RandomXYZb, Randomran, Remisebastien, Rettetast, RexBrynen, Rich Farmbrough, Rjwilmsi, Rlippert, Rlrr, Roberthu179, Rodo82, Roger.smith, SMcCandlish, Satish2106, Scarletb, Scoo, Scwlong, Sega381, Serge, Sergio.ballestrero, ShelfSkewed, SimonJETaylor, Simulation123simulation123, Slakr, Smith609, Sp3, Sphinx, Stack, StaticGull, Stephanhartmannde, Stephenb, Stephenchou0722, Stopitallready, Stricklandjs59, Tarotcards, TastyPoutine, Tawker, Tecknotrove, Tedickey, Telamon, The Anome, The Bushranger, The Thing That Should Not Be, The demiurge, The wub, ThreePD, Thseamon, Tiddly Tom, Tim.cooke2007, Truecobb, Tuckerj1976, Valueyou, Vanisaac, Vespristiano, VictorAnyakin, Vikom, Vsims, Vuo, Walter Görlitz, Wiki alf, WikiDao, Wikid77, WilfriedC, Will Beback Auto, Wisdom89, Woohookitty, Yeom0609, Ymoore33, Zaphod Beeblebrox, Zennie, Δ, 501 anonymous edits

Photography *Source*: http://en.wikipedia.org/w/index.php?title=Photography *Contributors*: 100110100, 11kseeker, 12ali34, 165.21.83.xxx, 1p2o3i, 2344fvdsfgdfsg, 27 Juni, 28421u2232nfenfcenc, 7castle, 7eden, 84user, 8ung3st, 93bairdy, 9tmaxr, A16wolfpup, A8UDI, ABF, AJCham, AVarchaeologist, AWR, Aaron Brenneman, Abdullais4u, Abstractjazz, Accurizer, Achristoffersen, Acroterion, AdRem, Adamskii, Adashiel, Addshore, Adoniscik, Aehlam, Aff123a, Ahernmas214, Ahoerstemeier, Aillema, Aitias, Ajcheema, Alan Liefting, Alansohn, Alborzagros, Aleksa Lukic, Alex Nadtoka, Alex Schult, Alex groos, Alexandermissen, Alexandrov, Alias Flood, Alistair.s, Alksub, Allartschools, Allstarecho, Altenmann, Alvinglori, Amazingflyingrabbit, Ameyjw, Amir beckham, Amphotog, Anagnam, Anastasios, Andre Engels, Andres rojas22, Andrew Steele, Andrewodom, Andrewpmk, Andy M. Wang, Andylimdotcom, Anekaindah, Angrysockhop, AnimAlu, Animum, Anna Lincoln, Anonymous Dissident, Antandrus, Anthonyasael, Anthonyhcole, Anthonymyre311, AntiVan, Antonio Lopez, Anupam9987, Apb, Aphaia, Appeltree1, AravindBalasubramaniam, Archeofoto, Arctic.gnome, Arolga, Arpingstone, Arpitt, Arrjae, Art LaPella, Artbunni, Arthur Rubin, ArthurSuna, Artincontext, Ashwaniwebacom, Asmah01, Atif.t2, Atulsnischal, Avenue, Avicennasis, AxelBoldt, Ayoub229, B. Nuhanen, BIsopp, BW, BabelStone, Backdropsmadhu, Badstoat, Baffle gab1978, Barek, Baronsen, Barras, Barticus88, Basz, Bayg, Bazzargh, Bcarman, Bdesham, Beardless, Beatlesone, Beetstra, Begoon, Bellyboy72000, Bemasher, Bergsten, Betterusername, Beyond My Ken, Bfigura's puppy, Bidiot, Biffer, Big puss, Bill Shirk, Billbouchard, Bills.books, Billsalter, BlaiseFEgan, Blog4u2, BlueDevil, BobTheTomato, Boba-Fett64, Bobjagendorf, Bobke, Bobo192, Bongwarrior, Bonsairolex, Bookofjude, Borislav, Bovlb, Brandon, Bremerenator, BrenDJ, Brentnow, Bricktop, Bright-young-thing1990, Brion VIBBER, BrokenSegue, Brunnock, Bruxism, Bryan Derksen, Buchangrant, Bunchofgrapes, Bundymas214, Bunnystars123me, Bus stop, C.anguschandler, CDA, CO, CQJ, Cactus.man, Calabe1992, Calieber, Caltas, Camembert, Can't sleep, clown will eat me, Canadian-Bacon, CanadianLinuxUser, Canderson7, Canjth, Captain panda, Carmichael, Catchytitles, CatherineMunro, Celain, Celarnor, Cenarium, Cflm001, Chamal N, Chameleon, CharlieCLC, Charmcculough, Charvex, Chasingsol, Chicboiii, Chochopk, Chocoforfriends, Choi229, Chris the speller, Chrisey91, Christopherlin, Chuck02, Chuenprayothmas214, Chzz, Ciceronl, Cintita, Claidheamohmor, Clappingsimon, Clark candace, Click23, Climbeverest, Closeapple, Closedmouth, Closenplay, Cmichael, Codwiki, Colin.nuke, CombatCraig, Cometstyles, Comex, Comidadehospital, CommonsDelinker, Computerjoe, Comyns Berkeley, Conscious, Conversion script, Cooletster, Coolman789, Coolrighthere2, Corti, Corvus cornix, Courcelles, Cpl Syx, Cramyourspam, Csalzman, Cunliffe214, CyberEliza, Cyberico, CygnetFlying, Cyktsui, D-Rock, DARTH SIDIOUS 2, DCEdwards1966, DJ Clayworth, DMacks, DVD R W, DaRaeMan, Daa89563, Dac04, Daler, Dan6hell66, Daniel, Daniel C. Boyer, DanielCD, DanielEng, Danlev, Darguz Parsilvan, Darkwind, Darth Panda, Darx9url, Daverivera, Davesta707, Davewho2, David Edgar, DavidDouthitt, DavidLevinson, Dawn Bard, Daytona2, Daytonarolexboston, Dbachmann, Deadcorpse, Deelkar, Deflagro, Deglr6328, Dekisugi, Delicious carbuncle, Delldot, Demonsmercy, Denee12345, Denisarona, Denisshepherd, Denna Haldane, DerHexer, Derek Ross, Dfeuer, Dgatto, Diafaye, Dicklyon, Diego UFCG, Digital1, Dingemansm, Dipics, Discospinster, Disdero, Disneycat, Dispenser, Djshaggy007, Dkroll2, Dlohcierekim, Dnynumberone, Doc glasgow, DocWatson42, Dori, Doulos Christos, Download, Dr.frankzappa, DragonflySixtyseven, Drahgo, Dreg743, Dricherby, Ds visualist, Dschwen, Dsw4, Duk, Dvyost, Dysepsion, E20Ci, ESCapade, EarnestyEternity, EbonyFleming, Ebooks, Eclecticology, Edison, Edward, Efghij, Egil, Elenseel, ElizabethFong, Ellmist, Ellybear5, EmadIV, Emilia0919199, EmilyDoyle, Enigma7seven, Enrichyourmind, Enviroboy, Epbr123, Equazcion, Erianna, Ericd, Erin.Bleure, ErinHowarth, Esanchez7587, Escape Orbit, EscapeTechwriter, EscapingLife, Eubulides, Eurekagray, Everyking, Ewawer, Ex nihil, Exlibris, Extraordinary Machine, FF2010, Falcon8765, Fan-1967, Farosdaughter, Fashionphotographer10, Favonian, FayssalF, Fcueto, FelisLeo, Felizdenovo, Fero.Mors, Fetchcomms, Fg2, Fieldday-sunday, Finlay McWalter, Fir0002, Firien, FisherQueen, Flamingspinach, Flash19901, Fletcher, Floydian, Flubeca, Flush752, Flyguy649, Flyhighplato, FlyingToaster, Focusnetwork, Fontboy, Fontgirl, Foogerbop, Fountains of Bryn Mawr, Fplay, Fraggle81, Fram, Franz Jantzen, Fredkamphues, Frood, Frozen4322, Funandtrvl, Fyrdahl, Fæ, GB fan, GWeaver, Gabbe, Gabrielone, Gadfium, Gaelen S., GageSkidmore, Gah4, Gaijin42, Gaius Cornelius, Galactor213, Galoubet, Garbagemania, Garland brandon, Gbleem, George dodds, Geoshrad, Germandemat, Ghijo12332, Giannivenice, Giftlite, Gillaura, Gilliam, Gioj50, Girolamo Savonarola, Gisle, Gogo Dodo, GoneAwayNowAndRetired, Gooberglobber, GordonMcKinney, Gotcheeze5793, Gphoto, Gpohara, Gracenotes, GraemeL, Grafen, Graft, Graham87, Grandcraft, Grayshi, GreatWhiteNortherner, Gregmasterson, Gross.ra, Grunt, Gsarwa, Gscshoyru, Gsmgm, Gtg204y, Guaka, Gudhka229, Gurch, Guthrun, Gwalla, Gökhan, H2g2bob, Haakon, Hadal, Haikupoet, Hallows AG, Hanamelissa, Hdt83, Hemia, Bo, Hencecolor, Henning, HenryLi, Heron, Hertz1888, HexaChord, Hilosoph, Hippotipus, Hmains, Hoo man, HorsePunchKid, Howcheng, Htgalante, Htrapp, Hu, Hu12, Huon, Hussnainmughal, Huzefahamid, Hyacinth, Hyperflux, I8munkies, Iamseawolf, IceUnshattered, Icebox, IchWeigereMich, Idiotbooks, Iknowsomethingudontknow, Illinois2011, Ilyarmas214, Imroy, Indon, Inferno, Lord of Penguins, Infrogmation, Ingolfson, Inkaroad, Internedko, Iridescent, Irrc irri, Ivan Teves, Ixfd64, J Milburn, J.delanoy, J991, JAF1970, JDGree916, JForget, JIMBEAN55, JONSTARFREAK, JaGa, Jack Merridew, Jackol, Jacobolus, Jacobp62, Jacopo188, Jagged 85, Jaimetout, Jamavr, JamesMLane, JamesNYC, Jamie Wade, JamieS93, Jamoche, Jan.Kamenicek, January, Jareha, Jauerback, Jean-Pol Grandmont, Jeff dean, JeffGowlett, Jeffmedkeff, Jengirl1988, Jenni gmas229, JeremyA, Jeronimo, Jessicasamara778, Jgritz, Jhendrixrbp, Jim Casper, Jim.henderson, JimmyModico, JinJian, Jjron, Jmartinhodesign, Jmencisom, Jmm629, Joe Rodgers, Joefromrandb, Joeymas, Joeywallace9, John254, JohnClarknew, Johnbibby, Johngrif, JohnnyDezyne, Johnnyswords, Jomama12, Jonadin93, Jorge Stolfi, Jose77, Joseph Solis in Australia, JoshuaNewman, Jowee, Jpbowen, Jpgordon, Jrockley, Jrpac, Jschnur, Julesd, Juliancolton, Jumbuck, Junglecat, Jusdafax, Just Another Dan, Just plain Bill, Justout09, Juzeris, Jwink3101, KC Panchal, Kablammo, Kadcpm88, Kaisershatner, Karebare89, Karen Johnson, Karen Joslin, KategibbsMAS226, Katesharp, Kavindu sivaraj, Kazubon, Kazudiaz, Kd4ttc, KeithH, Kelly Martin, Kelvinite, Kennygadams, Keonaprude, Khoikhoi, Killick, Kilonum, Kingboyk, Kingpin13, Kirrages, KiwiJeff, Kizor, Klausness, Klilidiplomus, Knight1993, KnowledgeOfSelf, Knucmo2, Knutux, Koliri, Kolkata culture, Koyaanis Qatsi, Kprateek88, Krashlandon, Kristephanie, Ktls09, Kturek, Kubigula, Kurt Eichenberger, Kurtdriver, Kuru, Kusunose, Kwamikagami, Kwekubo, Kwiki, Kylu, KyraVixen, Kzollman, LAX, La goutte de pluie, Laemokid, Lalphoto, Lammas214, Lander777, Largesock, Lauramelindajean, Laurascudder, Leandrod, Lechasseur, Leemas214, Leifern, LeoTran, Leon7, Leonard G., Levineps, Liam987, Lightmouse, Ligulem, Linkspamremover, Linuxlad, Linweizhen, Little guru, LittleOldMe, Llywrch, Logan, Logical2u, Lollies10, Lonewolfbme, Looxix, Lotje, LoveKiddoh, LovesMacs, Loveshorty, Lpsc007, LtPowers, Lucyin, Lugnuts, Luispaterna, Lukasinla, Luloz11, Luminiscence, Lumos3, Lunkwill, Lygor, MC MasterChef, MER-C, MLKLewis, Maarten4561, Macaddct1984, Macedonian, Mactographer, Macy, Magnus Manske, Malcolm Farmer, Man vyi, Mani1, Mannuak, Mannyram24, Manop, Marc703, MarcoTolo, Marcus334, Marielou, Marieluy143, Mark Arsten, Mark.murphy, MarkBrooks, MarkSutton, MarkSweep, Markb, Martarius, Martin Geisler, Martin451, Martinmaniac, Matthew Fennell, Matthew Gillis, Mayfly may fly, Mboverload, Mbz1, McGeddon, McSly, Mctrain, Mechanical digger, Meekywiki, Melderman, Mermaid from the Baltic Sea, Mfbz78, Michael

Ronayne, MichaelMaggs, Michaelas10, Midgley, Midsolo, Mike Rosoft, Miked918, Mild Bill Hiccup, Millasorel, Mindmatrix, Minimac's Clone, Minority2005, Miro modo, Mirv, Missshmu, Mjandrew, Mjpieters, Mmhood, Mobius, Modernist, Modulatum, Mohamedashraf0, Mominafrede, Mononomic, Montgomery '39, Moogwrench, Moonx, Mosaic guidance, Mossdavis, Motorrad-67, Mpk240, Mr Stephen, Mr. Good, MrOllie, Mrpoopsypanties, Mrshoestring, Muijz, Muiranec, Muzzy333, Mvwhitla, Mwanner, Mxn, Mygerardromance, Mynamesezzy, Myrabella, Myraedison, Müslimix, Naddy, Naidoomas214, Nakon, Naomichanart, Natc373, Natural Cut, NatureA16, NawlinWiki, Nayvik, Nburden, Ndkl, Neftin, NellieBly, Nemu, Neshatian, Netsnipe, NewEnglandYankee, Newzild, Nguyen214, Nicenjuicy, NicholasJones, Nick Number, Nihil, Nihilon, Nihilus, NikOly, Nikki88, Nireekshit, Niro5, Nitya Dharma, Nivix, Nixeagle, Noldoaran, Nono64, Nosol, NrDg, Nsaa, Ntse, Numbo3, Nuujinn, Nwbeeson, Nystrxz, Nyttend, Nzphotos, O00thunderhawk00o, ONEder Boy, Oblivious, Obradovic Goran, Obrienmas214, Oceans and oceans, Octave.H, Oda Mari, Ohnoitsjamie, Oicumayberight, OlEnglish, Oldcamera, Oli Filth, Olivier, Omar77, Omicronpersei8, Omnipaedista, Omshivaprakash, Omulazimoglu, OnePt618, OpDe, Optimist on the run, Osh716, Owen214, Oxymoron83, PPGMD, Palnu, Parhamr, Paris By Night, ParisianBlade, Parker229, Passion3545, Pat Kelso, Patlittle, Pattersonmas229, Paul Kelley, Paul Moscrop, Paul229, Paxse, Pchov, Pde, Performance Impressions, Peter.C, PeterSymonds, Peterlewis, Petersam, Pewwer42, Pfalstad, Pgarwood, Pgk, Phaldo, Philip Trueman, Phooto, Photo301, PhotoGuideLiz, Photojack50, Photoluc, Photomanjace, Phototubeone, Phydend, Piano non troppo, Picturetokyo, Pigman, Pimpinpunk, Pinethicket, Pingveno, Pink!Teen, Pinkadelica, Pinkville, Pit, Pixel8, Place 152, Pladask, PlatinumX, Pmlineditor, Postdlf, Ppinheiro, Prodego, Professor Fiendish, Projectblue, PseudoSudo, Pseudomonas, Ptk, Punkeychick7, Quadell, QuentinUK, Quest for Truth, Qwyrxian, Qxz, R'n'B, RHaworth, RJHall, RJN, RPSM, RaCha'ar, Rachdi, Rada, RadicalBender, RainbowOfLight, Ralarare, RaseaC, Ratarsed, Raven in Orbit, Rawling, Razorflame, Rcoble, Rdikeman, Rebornsoldier, Reconsider the static, Recury, RedCoat10, Redvers, Reedy, Reellis67, Rein213, RenniePet, Rentaferret, Research Method, Resourcesforlife, Retro00064, Rettetast, Revotfel, RexNL, ReyBrujo, Rhobite, Riana, Rich Farmbrough, Richardgay, Richm711, Rickjpelleg, Riggsaaron, Rivertorch, Riyonosuke, Rjwilmsi, Rmhermen, Robert Merkel, Robin klein, Rocksanddirt, RodC, Rodney Topor, Rokstr, Romanito, Roofbird, Rotundo, Roy da Vinci, Rror, Rsm99833, Rtfisher, Runesmith, S kander gul, S.raheja, SSJ 5, Sadiasalam, Salamurai, Sam Korn, Samsara, Samuel Grant, Sandy305, Sannse, Sansom Chris, Sarah Sandin, Sarathy wiki, SchulteMAS214, Schultzpaz, Scjessey, Scorpios, ScottJ, Scotty T4, Scotty T42, Scriberius, Sdkstudios, SeanMack, SeannaG, Seaphoto, Seattleart25, Secundus Zephyrus, Seglea, Seneviratnemas214, Seraphim, Seth Ilys, Sevenlinefeatures, Sgraziano, Shaakunthala, Shadow1, Shanes, Shaw Westminster, Shaykhspeara, Shell Kinney, Sheluvshoes, Shenme, Shewpao, Shirulashem, Shoeshirt, Shreshth91, Shubhamiitr, ShutterBugTrekker, Sicoactiva, SimonLyall, SimonMayer, SimonP, Sir Floyd, SirJello37, Siroxo, Skoobel, Slakr, Slazenger, Slo-mo, Slpayne, Slysplace, Smallmas214, Smartse, Smsarmad, SmthManly, Snigbrook, SoCalSuperEagle, SoWhy, Soccerking5052, Solipsist, SomeStranger, Soroosh.a, Soulrefrain, Spangspang, Sparkit, Spartan-James, Spectrojin, SpeedyGonsales, SpigotMap, Spiliotopoulou, Spitfire, Sreejithk2000, Srice13, Srleffler, Steeev, Stephan Bourgeois, Stephenb, SteveHopson, Steven J. Anderson, Stickee, Stui, Subversive.sound, SupFanat, Surrealcolor, Sweetbug17, Symane, Sypecom, Szaydel, TFMcQ, THEN WHO WAS PHONE?, TPK, Tangothellama, Tannin, Tarquin, Tarun2k, Tasou-azimuth, Taylormas229, TboneMN, Tdadamemd, TedE, Tedder, Tgreensuno, The Anome, The High Fin Sperm Whale, The Thing That Should Not Be, The Transhumanist, The Transhumanist (AWB), The undertow, The wub, The7thPython, TheDandyPhoto, TheKMan, TheMindsEye, TheStick, Thebestofall007, Thejerm, Theshibboleth, Thesmart177, Thingg, Thomas Blomberg, Thorseth, Thue, Tiberius47, Tide rolls, Tiffanyadamay, Tiggerjay, Tilamcarpets, Tim1965, Timoooo206, Tinanibbana, Tindozi, Tintenfischlein, Tinujohnmathews, Tiptoety, Titopao, Tobby72, Tokind, Tom Morris, TomS, Tomatoman, TonyClarke, Tonym88, Tooto, Tovisbratsburg, Trampledtown, TravisNygard, Traxs7, Trevor MacInnis, Triage, Triddle, Triona, Trusilver, Truth2244, Tsetsalullu, Tungsten235, Turian, TutterMouse, Twistmyarm, Tyghbnm, Tyrenius, Tyw7, Ugoalsugo, Ukkor, Ukt-zero, Ulpianus, Ultratomio, Uluvmeistarburst, Uncle G, Uncle Milty, Unclerodya, Unschool, Updatedinformation, User2004, Usethebumrag, Utkarsh.blackbelter, VQuakr, Valfontis, Vandriel1325, Vanished 6551232, Vanished user 39948282, Vanvisua, Veinor, Velella, Venturinik, Versageek, Versus22, Vipinhari, VivC, Vivalagloria111, Vrenator, Vsmith, WOSlinker, Waloitz, Wapcaplet, Warlordwolf, Waterfallsrus, WatermelonPotion, Wavelength, Wayne Slam, Wdpics, Weien, WereSpielChequers, Wik, Wiki-uk, WikiZorro, Wikibob, Wikipedian Penguin, WikipedianMarlith, Wikipelli, Wildfire1961, WillGuthrie14, William Avery, Wimt, Winjay, Wisco, WojPob, Wolfgobbler, Woodpainter, WpZurp, Wuffyz, Ww2censor, Wwagner, Wyss, X-x-blonde-x-x, X201, Xauxa, Xe0net, XjakusX, Xugglybug, Xugleee, Yakudza, Yamamoto Ichiro, Yanivg, Yankeemvp313, Yeshuaherenow, Yoenit, YourEyesOnly, Z david ross, Ze miguel, Zest777, Zondor, ZooFari, Zozza, Zundark, Zzuuzz, Zzyzx11, , 2391 anonymous edits

Visual_effects *Source*: http://en.wikipedia.org/w/index.php?title=Visual_effects *Contributors*: 51mfetic, ALoopingIcon, Abyssadventurer, Adeladili, Akjar13, Anshbruno1, Arevalo.Brian, Asydwaters, Aviseries, Binksternet, Bnewcomb, Bookuser, Bradv, Bwoloshyn, CGIExperte, Chris 42, ChrisGualtieri, Chrisz1, Ckatz, Classicfilms, Courcelles, Cowman109, DVD R W, Daquake, DavidFarmbrough, Deniznesrin, Discospinster, DreamGuy, E Wing, Echoray, Eekerz, Effstops, Egrabczewski, Elementxcreative, Evanjacobs, Fama Clamosa, Fitch, Freakofnurture, Furrykef, Gabbs1, GeoffPedder, Geoffmark, Giantstepsvfx, Ginesh, Girolamo Savonarola, Gmankiller555, Gogo Dodo, Greg lawrence, Grook Da Oger, HALVFX, Hede2000, Heightwatcher, Iggy35, Intgr, Iridescent, Izu81, JKWELL, Jagriff1, Jahshaka777, Janke, Jeanmarco12, Jeffokun, Jeremy Butler, Joe animator, Jotty, Jvecchia, Jweiss11, Katieh5584, Kimchi.sg, Kirill Lokshin, Kneelbeforezod, Kuru, Leonard^Bloom, Lindosland, Lookfx2, Lradrama, Lucky 6.9, Marasmusine, Marcd30319, MissHolly, Mohansen11, Monkeyproblem, Nachoman-au, Neonmarg, Nitinchand004, Ohnoitsjamie, OlEnglish, Pjoef, Prodego, Quietly, Quintote, RHaworth, Reaper Eternal, Reinyday, Renegademax, Reyk, Rocka122, Ronbrinkmann, Ronz, Samuel J. Howard, Shail.3d, Solipsist, Sputnik369, The Colclough, TheRealFennShysa, Tommyinla, Tregoweth, Tresemilita, Twiceuponatime, Tysto, Utilizer, ValterVB, Vanka5, Vfxblog, Wiki187, XiaoMiFeng, Xjosh2k6x, Xxpain, Yoichi3, YourEyesOnly, Z10x, Легат Ская, 200 anonymous edits

Slow_motion *Source*: http://en.wikipedia.org/w/index.php?title=Slow_motion *Contributors*: 16@r, Abridges01, Akeron, Armadillo7777, Asenine, Azucar, BarretB, Bhdrhntbk, CRJO-CRJO, Cantons-de-l'Est, Ckatz, Copana2002, Dewarsnsoda, Doctor Sunshine, Douglasr007, Eekerz, Eharney, Endie, Felipe P, Fred Bauder, Gaius Cornelius, Gogo Dodo, Gugganij, Gurch, Harami2000, HighSpeedWizzard, Imnotminkus, Inferno, Lord of Penguins, Ixfd64, JForget, JFreeman, Janke, KJS77, KayChase, Kimiko, Kjzhg, Koyaanis Qatsi, Lajbi, Lightmouse, LunarLander, MER-C, Mandarax, Mattgirling, Mcy jerry, Measly pawn, Meco, Moink, Mswake, NJM2010, Nakon, Nbarth, NeuronExMachina, Nick R, Ohnoitsjamie, OneSixOne, Ophois, Patriarch, Patrick, Philopp, Plowboylifestyle, Racula, Rebecca Bangay, Rich Farmbrough, Schmiteye, ShakingSpirit, Shammack, ShelfSkewed, SimonP, Slo-mo, TMC1982, Tad Lincoln, Tdadamemd, Telecineguy, The JPS, TheHYPO, Thehardwareman, Threegirlz, Tide rolls, Tukkaatje, Twinsday, Ubuntufish, Ungvichian, Vagrant, Wahkeenah, Wipe, WolfgangFaber, Woohookitty, Yarnalgo, Yojimbo501, ZeroOne, Zidonuke, 80 anonymous edits

Camera_angle *Source*: http://en.wikipedia.org/w/index.php?title=Camera_angle *Contributors*: Benjamin Pineau, BilCat, Egrabczewski, GHcool, George100, Girolamo Savonarola, Ground Zero, Hahalolz, ISTB351, Jnestorius, JordoCo, Khazar, Makeemlighter, Newsmare, Nick Number, Nsaa, PamD, Pjoef, R.P.Nambi, Rich Farmbrough, Serenatk, Shuipzv3, SimonP, TejasDiscipulus2, The Thing That Should Not Be, Thehelpfulone, Tricia Ziemer, Ukexpat, William Avery, Wknight94, Woohookitty, ישו דוד, 62 anonymous edits

Morphing *Source*: http://en.wikipedia.org/w/index.php?title=Morphing *Contributors*: 16@r, 5150pacer, Acidburn24m, Adam watson st wilfs, Addysoh2008, Adoniscik, Aetheling, Alexf, AliceJMarkham, Antandrus, April Arcus, Ark42, Arthur Frayn, Atakdoug, BenFrantzDale, Blainster, BlastOButter42, Bryan Derksen, BuickCenturyDriver, CanisRufus, CatherineMunro, Ccacsmss, Chocolateboy, Chris the speller, Closedmouth, CyclopsScott, D V S, DK4, Dah31, Dale Arnett, Daniel Mietchen, Danielfolsom, Darkwind, Davidhorman, Deadworm222, Dlohcierekim, E Wing, Elagatis, Etolpygo, EvilCouch, FleetCommand, Furrykef, Hannu, Infodriveway, Ink Runner, Intelligentsia, Itcleft, JNW, Jarred5211992, Jeevies, Jon Ascton, Joshr676, Jtuttle, JuJube, Karenjc, Kilo-Lima, Kingboyk, Korin, Kristoguy, Kvn8907, L Kensington, Lahiru k, Lethe, Lotje, M0rphzone, MER-C, Martarius, Martinevans123, Mayalld, Michael Snow, MikkoM, Minesweeper, MonkeyMumford, Mortedearthur, Noldoaran, Oicumayberight, PAR, PBP, Pcarbonn, Perfecto, Peter Kaminski, Piano non troppo, Pietron3000, Quocanh001, RJFJR, Racerx11, Ramon.cutanda, Robert Merkel, Rror, Scarce, SchfiftyThree, Scoty6776, ShelfSkewed, Sin-man, Sirgregmac, StaticGull, Stellmach, Steventx7, Stw, Tedfay, That Guy, From That Show!, Tisane, Tolandc, Ugly Bob, Uncle Dick, Warren, Wcpao, X!, Xanatos290, Yaminikanchi, Zack, ZooFari, Zrulli, आशिष भटनागर, 169 anonymous edits

Virtual_cinematography *Source*: http://en.wikipedia.org/w/index.php?title=Virtual_cinematography *Contributors*: Asetyyli, Brookie, Cmdrjameson, Dto, Ettrig, Girolamo Savonarola, Jdonati, Kollision, M jurrens, Machinima, Mahanga, Malcolma, Moe Epsilon, Pearle, Pharos, Piisami, Plowboylifestyle, RJFJR, Rich Farmbrough, Rjwilmsi, S, 16 anonymous edits

The Matrix *Source*: http://en.wikipedia.org/w/index.php?title=The_Matrix *Contributors*: *Ria777*, -Midorihana-, .:Ajvol:., 02gouldenr, 0bhimsingh, 16@r, 17Drew, 1ocus, 200.191.188.xxx, 2012WorldPeace, 21655, 23skidoo, 3wines, 4v4l0n42, 5 albert square, 5ko, A. C. Jakob, ARGOU, AVand, Aabelaros, Aaron Schulz, AaronCBurke, Aavindraa, AbsoluteGleek92, Abuk78, Access Denied, Acipsen, Adam2610, Adamcobb, Adashiel, Admeister200x, Adornix, Adraeus, Adzze, Aerowiki5667234, Afaprof01, Againme, Agentbla, Ahoerstemeier, Ailric, Aitias, Ajcfreak, Ajraddatz, Aktsu, Alai, Alakazam, Alan Canon, Alansohn, Alberteinhard, Aleoconamor, Alex, Alex 8194, Alex-thong, Alexjhe, Alfakim, Alientraveller, AlistairMcMillan, Alonelyend, Alphius, Alysion, Ameliorate!, AmiDaniel, Anab10sis, Anand011892, Anatidae, Anclation, Andonic, Andrewpmk, Andromine, Andru nl, Andrzejbanas, Andturim, Andycjp, AngeAnge22, AngelOfSadness, Angelic Wraith, Angemon 92, Angie Y., Angus Lepper, Animum, AnmaFinotera, Annabella Chronos, Ansible, Antaeus Feldspar, Antandrus, AnthonyQBachler, Anthonyxjames98, Antonio Lopez, Antriksh Yadav, Anville, Apollo13editer, Apox, Appraiser, AprilHare, Arag0rn, ArcAngel, Arcayne, Arctixfox, ArglebargleIV, Arjun01, Arpuns, Art LaPella, Artelove33, Artethical, Aseld, Ashmoo, Ashthecat3, Astral, Asylum 0, Atlan, Atrox4444, Auger Martel, Autonova, Ayrton Prost, Badgey, Bakken, Bakosy, Balsa10, Bardego, Bart133, Batman194, Bauani, Beckykaiser, Beeblebrox5000, Beeswaxcandle, Beetstra, Beier5, Belleedit, Bellpepper, Ben Ben, Ben D., Bencey, Bender235, BenjyBoy2000, Beno1000, BerelZ, Besidesamiracle, Betty Logan, Bevo, Beyond silence, Bhikipedia, Big Bird, BigCow, Bignole, Bijo.mala, Billybosky, Bkwillwm, Blair25, Blanchardb, Blatantcontent, BlazerKnight, Blue520, Bluerasberry, Bmicomp, Bob Ilyani, Bobet, Bobo192, Bobsponge, Boneheadmx, Bongwarrior, Bonus Onus, Bovineboy2008, Boyton, Brandeks, Brandon, Brianga, Brink9, BrokenSegue, BulletBillTime, Bumba claat, Burgerman277, Butros, Byelf2007, ByrnedHead, C-squared, CALR, CaloCaleta, Calvin 1998, Cameron Nedland, Can't sleep, clown will eat me, Canley, Canoftuna, Cantus, Capricorn42, Captain Proton, Cardinal Wurzel, CardinalDan, Carlosguitar, Carlosp420, Carlsmith, Carmichael, Carolaman, Cate, Catgut, Catticusratticus, Causa sui, Cburnett, Cercia, Cesoid, Chad625, Chairman S., Cheesemeister, Chowbok, ChrisGriswold, ChrisGualtieri, ChrisU, Chrislk02, Christopher Parham, Christopherlin, Cielomobile, CillanXC, Cirt, Ckatz, Clashfrankcastle, Closedmouth, CmdrClow, CollisionCourse, Comalies, Cometstyles, Commander Keane, Comp25, CompIsMyRx, Conf, Conkerhe, Conman71, Conti, ContiAWB, Conversion script, Coolaries4ever, Coredesat, Courcelles, CowboySpartan, Cp111, Crabula, Craigy144, Crazynas, Crotchety Old Man, Crserrano, Cryptic, CryptoDerk, Cst17, Curps, CyberSkull, Cyberchimp, Cyrius, DARTH SIDIOUS 2, DDerby, DHeyward, DJ Clayworth, DJAftermath, DMCer, DMacks, DRosenbach, DT29, DVD R W, DXBari, Da Vynci, Daa89563, Dale Arnett, Damian Yerrick, Daniel C. Boyer, Daniel Olsen, DanielJosphXhan, Darikrox7, Darkness2005, Darkwarriorblake, Darrelljon, Darrenhusted, Dave6, Davewild, David Gerard, David Hochron, DavidDW, Davidf2, Dbfirs, Dchall1, DeLarge, DeWaine, Dead3y3, DeadEyeArrow, DeathSeeker, Deineka, Dejvid, Delfuego, Delldot, Demoman87, Denial, Denimadept, Dennis Bratland, DennisTheTiger, Denor, DerHexer, Devilmaycares, Dgw, Dheiney, Dickytheorienteer, Didgeman, Diggerjohn111, Dillonwaldo, Dim386, Dimadick, Dinoguy1000, Dipakgoyal, Discospinster, Dismas, Diza, Doctor who9393, DoctorWho42, Doctorwhofan2008, Domharrington, Domthedude001, Donmike10, Donny, Doppelgangland, Doughboy, Douglas.kastle, Dposse, Draconis Karite, DragonflySixtyseven, Dreadstar, DriveMySol, Drunkenpeter99, Dsine, Dstanton uf, DuncanWilkie, Dustimagic, Duttler, Dwarf Kirlston, Dynaflow, Dynesclan, Dysprosia, E Pluribus Anthony, E Pluribus Anthony redux, EEMIV, EWS23, EZio, EamonnPKeane, Eataubm, Echuck215, Eddie.willers, Edisonqv, Edokter, Edward, Edward Turner, Edward Z. Yang, Eekerz, Ekabhishek, El Mayimbe, Elbanana, Elcapitane, Eleassar777, EliasAlucard, Eloquence, Eluchil, Eman007, Emma -Babe, Enfestid, Ensignyu, Enzo Aquarius, Epbr123, Epiphany11, Ericl234, Erik, Erik Garrison,

Erik9, Esperant, Eternal Elder, Etr, Everyking, Evil saltine, Evlekis, Explicit, Exploitation, Extracreditassignment, FF2010, FLStyle, FMAFan1990, FaithLehaneTheVampireSlayer, Fakefix, Falcon8765, Fama Clamosa, Fan-freaken-tasic, Fang Aili, Fantasy, Faradn, Farmercarlos, Fastfission, Fastily, Fataltourist, Feitclub, FentonGlass, Fer Shi, Ffaker, FiGhT 12, FiP, Fibonacci, Figure, Fish and karate, FisherQueen, Flewis, Flin07, Flockmeal, Floydgeo, Flyguy649, Fnfd, Formeruser-81, Fourthords, Fr33kman, Fractalgfx, Frank Romein, FrankCostanza, FrankRizzo2006, Franklinpierce3, Freakofnurture, FreshOnTheScene, Froid, Fruta995, Frymaster, Fsiler, Fuckmatrix, Fuhghettaboutit, Furrykef, Fwachowski, Fæ, Fëaluinix, GHcool, GNozaki, Gadfium, Gaily Ezer, Gaius Cornelius, Gallaghp, Gamaliel, Gameboy1993, Gareth McCaughan, Gargabookofayr, Garion96, Garveybjr, Gblaz, Gcanyon, Geekdiva, Gemtiger, Geni, George Ho, George100, GeorgeOrr, Ghamer01, GimliDotNet, Gjd001, Glane23, Glst2, Gnostril, GoHawks4, Goblin, Godaiger, Godlord2, Goethean, Gogo Dodo, GoingBatty, GoldenGoose100, Gonnym, Goon Noot, GorillaWarfare, GraemeL, Grafen, Graham87, Grahamdubya, Graveenib, Greengobbie92, Greenguy1090, Grendelkhan, Greswik, Grstain, Gsmgm, Guiltyspark343, H2g2bob, Hackwrench, Hadal, Haggard, Hairy Dude, Hall Monitor, Hamiltonstone, Hannes Karnoefel, Haon 2.0, HappyInGeneral, Happykafka, Hari, Haritada, Harkenbane, Harry, Haseo9999, Haza-w, Headwes, Hearfourmewesique, Hearts101, Hectorhector, Hede2000, Heilemann, Hello32020, HelloThereNow, Hemanshu, Henryjake121, Hephaestos, Hershey311, Heyelle, Hfastedge, Hgriggs, Hideonakata, Highonbread, Hiphats, Histrion, Hmrox, Hmwith, Homestarwunna, Hoovernj, Horologium, HotSauceHustla, Houdin654jeff, Htews, Hu, Hu12, Hui sk, Hullaballoo Wolfowitz, Husond, Hut 8.5, Hyacinth, Hydrogen Iodide, Hyliad, I-luv-the-navy, IRKAIN, IRT.BMT.IND, Ian Pitchford, Ice Vision, Iitmsriram, Ikh, IllaZilla, Ilovepie316, Immblueversion, Imroy, InShaneee, Infinium, Infocidal, Iondream, Iridescent, Irish Souffle, Islamisgr88, Ithinkhelikesit, Ixfd64, J.delanoy, J21, JAltman752, JForget, JTBX, Jacek Kendysz, Jackdaniels63, Jagged 85, JakeWharton, JamesBWatson, Jason M, Jason One, JasonAQuest, JasonLangan, Jauerback, Javascap, JavierMC, JayKeaton, Jayjg, Jbailo, Jeandré du Toit, Jeeves1618, Jeff G., Jeff3000, Jeffro77, Jemarre87, Jennavecia, Jennica, Jerem43, Jeremy Avalon, Jesuslikespancakes, JethroElfman, Jetro, Jgm, Jhsounds, Ji30i30, Jim1138, Jleedev, Jnc, Jncraton, Jni, Joakim Ziegler, JoanneB, Job11795, JobSmee, JoeBlowfromKokomo, Joebengo, Joelholdsworth, Joerite, John KB, John gaeta, John of Reading, Johnfn, Johnpseudo, Jojhutton, Jokestress, Jon Harald Søby, JonHarder, Jonas Viper, Jonathunder, Jonnabuz, Jookbox, Jordancelticsfan, Jordon Roy Gowans, Jorgicio, JoshuaZ, Jossi, Joy, Joyous!, Jpiece, Jrockley, JuJube, Judgesurreal777, Juggleandhope, Juhachi, Juliancolton, Jusdafax, JustAGal, Jwaycuilis, Jwdoom, Jwk3, Jzummak, KLLvr283, Kai Ojima, Karada, Karmagic, Kascnef82, Kasreyn, Kaylorific, Kbdank71, Kchishol1970, Ke6bus, Keilana, Kelly Martin, KerathFreeman, Kernow, Kerotan, Kether83, Kevinfergs, Kevinpaulscarrott, Keyser Söze, Khaosworks, KickAir8P~, Kidkeoni, Kidlittle, KieferSkunk, Kieff, Kieranthompson, Killingthedream, Kim Bruning, King nothing, King of the Dancehall, KingKane, Kingcole64, Kjd, Kjgmusic, Knnknn, KnowledgeOfSelf, Koavf, Kollision, Konman72, Kookyunii, Koveras, KozalBCE, Kripkensteiner, Krishvanth, Krogu, Kugamazog, Kumarpradheep, KungfuJoe1110, Kuru, Kyng, Kyrex, Kyz, L337 kybldmstr, LGagnon, Lacrimosus, Lamontacranston, Lavenderx, Ld100, Leader Vladimir, Leafyplant, LeaveSleaves, Lenaic, Lenin and McCarthy, Leussk, Levin, Levineps, Lg16spears, Liam Skoda, LifeStar, Lightmouse, LilHelpa, Lima Neto, LineOfBlackCars, Ling.Nut, Liquidtenmillion, Lisawong, Little sawyer, LittleDan, LittleOldMe, Live-info, Livewire1015, Lkesteloot, Llavigne, Lockeownzj00, Logixoul, Logotu, Lokantha, Lolnostanku, Loonymonkey, Lord Phillock, Loremaster, Loren.wilton, Loser578, Lowryba, Lošmi, LtNOWIS, Lucy8297, Lugnuts, Luigi30, Luigibob, Luna Santin, Lupin, M3taphysical, MATRIX, MECU, MER-C, MF-Warburg, MFNickster, MONGO, Maaajid, Macempty, Machine Elf 1735, Madchester, Maelfyn, Magioladitis, MajorB, Malaseb, Malcolm rowe, Malevious, Malkinann, Mallanox, Mandarax, Mange01, Mankar Camoran, Mansmokingacigar, Manticore, Manwithbrisk, Marcus Brute, Mark5677, MarnetteD, Martarius, Martin Ulfvik, MartinHarper, Master Deusoma, Master Jay, Matariel, Matilda, Matt Gies, Matt.smart, Matthew Proctor, Matthew Yeager, Mattingly23, Matzthebomb, Maury Markowitz, Mav, Max314, Maximus Rex, MaximusBrood, Mboverload, Meekywiki, Meelar, Meisterkoch, Melongrower, Menchi, Mervynvk, Metafact, Metao, Meursault2004, Mgiganteus1, Mhking, MiShogun, Mice, Michael Hardy, Michaelsbll, Mickyfitz13, Midgley, Mifter, Miguel.mateo, Mika1h, Mikael642, Mike Payne, Mike18xx, MikeLynch, MikeWazowski, Mikker, Millahnna, Millerdaca, Mimzy1990, Minesweeper, Mipadi, Mista-X, Misterandersen, Mkweise, Mnavon, Moink, Mojo2132, Moki80, Momus, Monkeymanman, Monkfishbandana, Monobi, Monoecus, Mookydowjones, Moreschi, Morgan Leigh, Moritheil, Mortyman2407, MouseLite, Moviecrazymattg, Moxy, Mpeisenbr, Mr. Billion, MrChupon, MrMPS, MrOllie, Mrshaba, Mschlindwein, Mspraveen, Mtcv, MuZemike, Muchmusic64, Mudel, Muhammad Hamza, Muhandes, Munci, Musical Linguist, Mwr99, My76Strat, Mysekurity, Mytildebang, NTox, Nadianadhirrah, Nafi1122, Nakon, Nalvage, Napolion, Naraht, Nasnema, Natalie Erin, Natalya, NathanHurst, NavarroJ, NawlinWiki, Nccsa186, Ncrfgs, Ndjokovic, Ndru01, Neander7hal, Nectarflowed, Neelix, NeiNie, Neilc, Nekrogeist, Neodammerung, Neonecho, Neowhat, Nesa0treb, Nescio, Netoholic, New Jack Swing, Nfichman, Nick R, Nicklasr3, Nickpetrellafilms, Nigelwalker, Nihilozero, Nihiltres, Nikai, Ninsheart, Nippoo, Nirvana2013, Nissir, Niwi3, Nixdorf, Niz, Nneonneo, Noldoaran, NorthernThunder, NotAnonymous0, Notmicro, Nowster, Numb094, Nv8200p, Nyxaus, OBloodyHell, OOODDD, Ocaasi, Oh babe, Olathe, Oleg Alexandrov, Olivier, OlofE, Omega214, Omicronpersei8, Omikron23, One, One of them, Oneiros, Onlyoneleft, Optichan, Orange Suede Sofa, Orangemike, Orelstrigo, Ori Redler, PYLrulz, Padraic, PadraigT, Panser Born, Pantothenate, Paranoid, Party29, Pashmina, Paul A, Paul Stansifer, PaulFCB, PauloIapetus, Pb30, Peabody80, Pearle, Pedro cunha, Pegship, Pejmany, Pentathor, Percevalles, Petegrew69, Pfalstad, Phbasketball6, Phil McMullin, Philip Trueman, Philwelch, Phthoggos, Piano non troppo, Picaroon, Pictureuploader, Pilotguy, Pjrich, Plasticup, Plrk, Podria104, Polisher of Cobwebs, Polly, Polyamorph, Potato dude42, Prem555, Premeditated Chaos, President Rhapsody, Proctris, ProtocolOH, PseudoSudo, Psy guy, Punk Rocker, PussyFart10000, Qtyo233, Quiddity, Quinnuendo, QuiteUnusual, Qwyrxian, REN, RJFJR, RK, RODERICKMOLASAR, Rachel0898, Ranchanak, Random User 937494, Ranjithsutari, Rat at WikiFur, Raven4x4x, Ravenmasterq, Ray andrew, Rayghost, Rbarreira, Rdl27, Reaper Eternal, Reaper X, Rebrane, RedHillian, RedWingedEagle, RedWolf, RedWordSmith, Redhaven27, Reedbeta, ReelAngelGirl, Regi999, Rejectwater, Renegader, RepublicanJacobite, RevRagnarok, Revan4000, Revth, Rfc1394, Rich Farmbrough, Richard ruffian, Richardcavell, Rickdrew, Ricky540, Ricky81682, Rjwilmsi, Rkb669, RlyehRising, Rmhermen, Roadrunner, RobJ1981, Robert Merkel, RobertG, Robertb-dc, Roberth Edberg, Robinh, Rockfang, Rockstone35, Rodhullandemu, Roller Coaster Fan, Roux, RoyBatty42, Rumiam, Rusted AutoParts, Ryoutou, S h i v a (Visnu), SCZenz, SD6-Agent, SISLEY, SJP, SQL Oracle, SS, SS451, SSkullcrusher, SajmonDK, Sam42, Samboy, Sammy&PJ, Sanius, Santiparam, Sartak, Satori Son, Sazerac, Sceptre, Schneelocke, SchuminWeb, Schwenkstar, Scooter2536, Scorpion0422, Scottlevy01, ScramJones, Sean D Martin, Seaphoto, Secretlondon, Sexybeast625, ShadowMan1od, ShadowOfEclipse, Shadowcaster187, Shadowjams, ShakingSpirit, Shanes, Shawnc, Shenme, Sherbertpi9, Shinerunner, Shyam29, Sietse, Silent Tom, Sionus, SirGrant, SirTopumHat, Sirjulio, Sitethief, Sjones23, Skakkle, Skarebo, Skier Dude, Skittleys, Skomorokh, Sky Attacker, Slackticus, Slady, Slakr, SlayBelle, Slicknick114, SlimVirgin, Sloane, Smelialichu, Smiloid, Smoothy, Smurfy, Sn0wflake, Snealri, SoSaysChappy, Soccerman90, Sockatume, Someguy1221, Somniloquy, Sonam8311, Sonicsuns, Sopoforic, Soren121, Sourcecode, Sparky the Seventh Chaos, Sparkzy, Spebudmak, SpeedyGonsales, Spektre0, Spellcast, Spencerian, SpinnakerMagic, Spra, Spriteless, SpuriousQ, Squawk7984, Srstudios123, Starwarsroxyeah, Statalyzer, Stentie, Steve, SteveCoppock, Steveosteve87, StormXor, Straightdude1, StreborKram, StringRay, StuRat, Stuclark33, Stuorguk, Sublime Satire, Suburbanslice, Sudharsansn, Sugar Bear, Sugarfish, SumnerRichter, Superm401, Supertigerman, Sweetfreek, Sweetlew13, Sycthos, TITandNICK, TS08707, Tahiti Charras, Tamara400, Tamfang, Tantek, Taoster, Taroaldo, Tarquin, Tasc, Tbhotch, Tdude418, TeaDrinker, Ted87, Teethmonkey, Teh tennisman, Teknorat23, Template namespace initialisation script, Terrasweet, Tesseract2, TexasAndroid, Texture, ThanosMadTitan23, The 80s chick, The Anome, The Crawling Chaos, The Cunctator, The Emo Panda, The Filmaker, The Giant Puffin, The Greatest Show On Earth, The Inedible Bulk, The Moving Finger Writes, The Rambling Man, The Thing That Should Not Be, The Well Wisher, The Wookieepedian, The ansible, The undertow, The wub, TheJC, TheMovieBuff, TheRealFennShysa, The_ansible, Thecheesykid, Thematrix15602, Thematrixhasyou, Themovementlives, Theorbtwo, There is no spoon., Thesonofabob, Thexile, TheyWalkAmongUs, ThinkEnemies, Thirdvertigo, Thiseye, Thismightbezach, Thrave, Tide rolls, Tiger Trek, Tigger69, Tilla (usurped), Tim from Leeds, Timotheus Canens II, TinyMark, Tired time, Tjmac23, Tjmage1, Tnxman307, Tom, Tom Lennox, Tommy11111, Tommy2010, Tommyt, Tony Sidaway, Tonymyles, Toresbe, Totorotroll, Trans4mers, Treaduse, Trevor MacInnis, TreyGeek, Triona, TrojanMan, Troy 07, Trusilver, Truthflux, Tstormcandy, Turnstep, Tuspm, Tutmosis, Txinviolet, Tylerwillis, Typhoon966, Typwriter92, Tømbs, Uker, Ukexpat, Ultimatemikito, UnR3al man, Uncle Dick, Unidef, UnitedStatesian, Universe, Updatehelper, Urhixidur, Useight, UtherSRG, Vadmium, Vanished user 39948282, Vanka5, Varco, Vaughnstull, VdSV9, Vejet, Vendettax, Verbose, VeryVerily, Vex5, VijayPadiyar, Vina, Vinhtantran, Violetriga, Viridian, Viriditas, Visor, Voidxor, VolatileChemical, Vyn, WBardwin, WCityMike, WLU, Wachstropfen, Wadems, Warpozio, Wayne Slam, Wayne317, Weplayit, Wesley, West.andrew.g, Weyes, WhisperToMe, Who, Whoisthecatalyst, Wifione, WikiLaurent, Wikibob, Wikimaster97, WikipedianMarlith, Wikipelli, Wikiteer461, Wikiwikifast, Willkm, Willthrillkill, Wilson44691, Wisdom89, Wjejskenewr, Wknight94, Wolfe604, WolfenSilva, Woodsstock, Woohookitty, Wow3, Wtmitchell, Wwagner, Wyatt915, XJamRastafire, Xaosflux, Xavcam, Xdcdx, Xenofan 29A, Xenophon777, Xeon25, Xezbeth, Xiaop, Xino, Xjaymanx, Xklsv, Xx ArBiT xX, Y0u, Y2kcrazyjoker4, YUL89YYZ, Yamakiri, Yamamoto Ichiro, Yamla, Yar Kramer, Ylbissop, Yomexzo, Yuri dragon 17, Yuval Y, Z3u2, Zepheriah, Zepheus, Zero1328, ZimZalaBim, Zoe, Zoop, ZroMan0, Zzuuzz, Zzzzz, 2522 anonymous edits

Still_camera *Source*: http://en.wikipedia.org/w/index.php?title=Still_camera *Contributors*: A2Kafir, Ahmed Elnagar, CanisRufus, Eep², Egil, Ericd, Everyking, Funandtrvl, Fæ, Hmrox, Hooperbloob, J Di, Kjkolb, Michael Hardy, Nescio, Twinsday, Walter Federico Graf, 9 anonymous edits

Mature_technology *Source*: http://en.wikipedia.org/w/index.php?title=Mature_technology *Contributors*: Alansohn, BeerBellyBob, Chocoforfriends, Dragon 280, Hit bull, win steak, Ingolfson, Kvn8907, Lamro, Matthew Yeager, Maxim, Mdd, TWCarlson, 8 anonymous edits

Image Sources, Licenses and Contributors

Image:Muybridge horse jumping.jpg *Source*: http://en.wikipedia.org/w/index.php?title=File:Muybridge_horse_jumping.jpg *License*: unknown *Contributors*: Fastfission, G.dallorto, MichaelMaggs, NYKevin, Oldstuff, Pitke, Wst

File:Horse simulator WWI.jpg *Source*: http://en.wikipedia.org/w/index.php?title=File:Horse_simulator_WWI.jpg *License*: unknown *Contributors*: User:Scoo

File:Christer Fuglesang underwater EVA simulation for STS-116.jpg *Source*: http://en.wikipedia.org/w/index.php?title=File:Christer_Fuglesang_underwater_EVA_simulation_for_STS-116.jpg *License*: unknown *Contributors*: NASA

File:lambda2 scherschicht.png *Source*: http://en.wikipedia.org/w/index.php?title=File:Lambda2_scherschicht.png *License*: unknown *Contributors*: Andreas Babucke

File:Simuladormotocicleta.jpg *Source*: http://en.wikipedia.org/w/index.php?title=File:Simuladormotocicleta.jpg *License*: unknown *Contributors*: User:Andrevruas

File:3DiTeams percuss chest.JPG *Source*: http://en.wikipedia.org/w/index.php?title=File:3DiTeams_percuss_chest.JPG *License*: unknown *Contributors*: Gene Hobbs

File:Vehicle simulator.jpg *Source*: http://en.wikipedia.org/w/index.php?title=File:Vehicle_simulator.jpg *License*: unknown *Contributors*: Bapho, Mattes, Nv8200p, Trockennasenaffe

File:ugs-nx-5-engine-airflow-simulation.jpg *Source*: http://en.wikipedia.org/w/index.php?title=File:Ugs-nx-5-engine-airflow-simulation.jpg *License*: unknown *Contributors*: Freeformer, Normsch

File:KSCFiringroom1.jpg *Source*: http://en.wikipedia.org/w/index.php?title=File:KSCFiringroom1.jpg *License*: unknown *Contributors*: NASA

File:Large format camera lens.jpg *Source*: http://en.wikipedia.org/w/index.php?title=File:Large_format_camera_lens.jpg *License*: unknown *Contributors*: User:Joanjoc

File:Contax-s.jpg *Source*: http://en.wikipedia.org/w/index.php?title=File:Contax-s.jpg *License*: unknown *Contributors*: Original uploader was Jeff dean at en.wikipedia

Image:Nikon-F2AS.jpg *Source*: http://en.wikipedia.org/w/index.php?title=File:Nikon-F2AS.jpg *License*: unknown *Contributors*: User:Moonx

File:Nikonf.jpg *Source*: http://en.wikipedia.org/w/index.php?title=File:Nikonf.jpg *License*: unknown *Contributors*: Original uploader was Jeff dean at en.wikipedia

File:minoxA.jpg *Source*: http://en.wikipedia.org/w/index.php?title=File:MinoxA.jpg *License*: unknown *Contributors*: Original uploader was Olegvolk at en.wikipedia

File:Leica-D-Lux-2-p1030299.jpg *Source*: http://en.wikipedia.org/w/index.php?title=File:Leica-D-Lux-2-p1030299.jpg *License*: unknown *Contributors*: User:Rama

File:Rob McArthur.jpg *Source*: http://en.wikipedia.org/w/index.php?title=File:Rob_McArthur.jpg *License*: unknown *Contributors*: FlickrLickr, FlickreviewR, G.dallorto, Infrogmation, Leoboudv, Para

File:Dark room.jpg *Source*: http://en.wikipedia.org/w/index.php?title=File:Dark_room.jpg *License*: unknown *Contributors*: User:Inkaroad

File:Nicéphore Niépce Oldest Photograph 1825.jpg *Source*: http://en.wikipedia.org/w/index.php?title=File:Nicéphore_Niépce_Oldest_Photograph_1825.jpg *License*: unknown *Contributors*: Nicéphore Niépce

Image:Latticed window at lacock abbey 1835.jpg *Source*: http://en.wikipedia.org/w/index.php?title=File:Latticed_window_at_lacock_abbey_1835.jpg *License*: unknown *Contributors*: William Fox Talbot

Image:Daguerreotype tintype photographer model studio table brady stand cast iron portrait photos.jpg *Source*: http://en.wikipedia.org/w/index.php?title=File:Daguerreotype_tintype_photographer_model_studio_table_brady_stand_cast_iron_portrait_photos.jpg *License*: unknown *Contributors*: User:Cramyourspam

File:Prokudin-Gorskii-12.jpg *Source*: http://en.wikipedia.org/w/index.php?title=File:Prokudin-Gorskii-12.jpg *License*: unknown *Contributors*: , digital rendering for the Library of Congress by Walter Frankhauser / WalterStudio

File:Photography using Canon Digital IXUS 850 IS.jpg *Source*: http://en.wikipedia.org/w/index.php?title=File:Photography_using_Canon_Digital_IXUS_850_IS.jpg *License*: unknown *Contributors*: Martinmaniac at en.wikipedia

File:Capas-d1.jpg *Source*: http://en.wikipedia.org/w/index.php?title=File:Capas-d1.jpg *License*: unknown *Contributors*: Original uploader was WikiCapa at en.wikipedia

File:Coolscan-V.jpg *Source*: http://en.wikipedia.org/w/index.php?title=File:Coolscan-V.jpg *License*: unknown *Contributors*: Original uploader was Jeff dean at en.wikipedia

File:K800i-back.jpg *Source*: http://en.wikipedia.org/w/index.php?title=File:K800i-back.jpg *License*: unknown *Contributors*: Asim18, MB-one, 2 anonymous edits

File:Freak Out, Oblivion, night.jpg *Source*: http://en.wikipedia.org/w/index.php?title=File:Freak_Out,_Oblivion,_night.jpg *License*: unknown *Contributors*: Freddo, JackyR, Man vyi, Solipsist, Takeaway, 1 anonymous edits

File:All Around Chajnantor.jpg *Source*: http://en.wikipedia.org/w/index.php?title=File:All_Around_Chajnantor.jpg *License*: unknown *Contributors*: Jmencisom

File:The Steerage 1907 Stieglitz Corrected.jpg *Source*: http://en.wikipedia.org/w/index.php?title=File:The_Steerage_1907_Stieglitz_Corrected.jpg *License*: unknown *Contributors*: SteveHopson

Image:Wootton bridge.jpg *Source*: http://en.wikipedia.org/w/index.php?title=File:Wootton_bridge.jpg *License*: unknown *Contributors*: Unknown photographer

File:Wrightflyer highres.jpg *Source*: http://en.wikipedia.org/w/index.php?title=File:Wrightflyer_highres.jpg *License*: unknown *Contributors*: Attributed to Wilbur Wright (1867–1912) and/or Orville Wright (1871–1948). Orville Wright preset the camera and had John T. Daniels squeeze the rubber bulb, tripping the shutter.

Image:OvercrankingTimeline.png *Source*: http://en.wikipedia.org/w/index.php?title=File:OvercrankingTimeline.png *License*: unknown *Contributors*: User:Plowboylifestyle

Image:DigitalSlowmotionTimeline.png *Source*: http://en.wikipedia.org/w/index.php?title=File:DigitalSlowmotionTimeline.png *License*: unknown *Contributors*: Plowboylifestyle

Image:HS-100-deck-ampex.jpg *Source*: http://en.wikipedia.org/w/index.php?title=File:HS-100-deck-ampex.jpg *License*: unknown *Contributors*: Telecineguy

File:Family Group.jpg *Source*: http://en.wikipedia.org/w/index.php?title=File:Family_Group.jpg *License*: unknown *Contributors*: TejasDiscipulus2

Image:Bush-Arnie-morph.jpg *Source*: http://en.wikipedia.org/w/index.php?title=File:Bush-Arnie-morph.jpg *License*: unknown *Contributors*: User:Iainf

file:The Matrix Poster.jpg *Source*: http://en.wikipedia.org/w/index.php?title=File:The_Matrix_Poster.jpg *License*: unknown *Contributors*: © 1999 Warner Bros. Entertainment Inc. (US, Canada, Bahamas and Bermuda); © 1999 Village Roadshow Films Limited. (All Other Territories)

File:LobbySceneScreenshot.jpg *Source*: http://en.wikipedia.org/w/index.php?title=File:LobbySceneScreenshot.jpg *License*: unknown *Contributors*: Autonova, Feydey, Mechamind90, Melesse, 2 anonymous edits

MIX
Papier aus verantwortungsvollen Quellen
Paper from responsible sources
FSC® C105338

Printed by Books on Demand GmbH, Norderstedt / Germany